£3.50

£3.50

Lullaby Wisdom

The Stories, the Songs, and the *Science of Soothing*

To hear the lullabies sung (usually in the informal, quieting, intimate voice styles of the caregivers who use them), use this link:

bit.ly/LullabyWisdom

(No spaces, and the L and W are upper case.)

This should take you to a page with all the audio files. Select one and three vertical dots may appear near the top right of the page. Click those and choose **"Download"**. Find the file in Downloads on your device and click on it to play it.

Collected & Edited by Licia Claire Seaman

Watercolor Illustrations by Peggy Dressel

Hardcover: ISBN 978-0-578-31613-0
Paperback: ISBN 978-0-578-31614-7

Cover design by Joshua Seaman
& Dietrich Seaman

Published by
Peaceful Pear Publishing
2715 NE 14th Ave
Portland, OR 97212

♪*A Note At The Beginning*

*T*ake me back to my beginnings with song. The cream-colored Mercury sedan held all ten of us in the two seats. The 'big three' sat on the back seat. Each held a 'War Baby'. Born during World War II, the 'War Babies' were no longer babies but the middle three's 'War Baby' nickname stuck throughout their lives. Bill held Betsy on his lap at the window seat. Helen held Molly with their feet on the hump. Nancy held Babs (that's six.) In the front seat, Mother held baby Bobby. I sat in the middle, and Daddy with his arms swirling over each other, wrestled the non-power steering wheel to drive the car up the curvy Green Hills roads on the crest of Portland, Oregon.

On the long trip up the Columbia River Gorge to visit Daddy's brothers in The Dalles, we would admire the welcoming basalt cliffs furred with forest, and waterfalls trickling or gushing down, leaping off the cliffs and bouncing on their way to the massive river below. The Columbia, widest in North America, always flowed on mystery and power.

We loved river grays, cliff browns, tree greens, sky blues, cloud and waterfall whites in silence ….sometimes. We were close like caramel corn. If talk turned picky or ugly, Daddy would fill up with an inhalation, then sound a song, *"The Same Silver Moon….."* and we all jumped in, "shining down through the trees" and on through at least two repeats of the whole song in harmonies of the 1930's. Then on to *"Shine On Harvest Moon"*, *"Down By the Old Mill Stream"*, *"Found A Peanut"*, *"Be Kind to Your fair-feathered Friends, for a duck can be somebody's father…."*, and more, until we got to *"99 Bottles of Beer"*. You get the point. The sonorous car went on and on, way after car radio reception was nil from entering rural cliff-ed Oregon.

My oldest sister Nancy, in her older years, reminisced,

> *"The Same Silver Moon" - Dad would sing that song when each baby was born. He would waltz with that baby, and hold the fussy baby and sing in a gentle timbre, "The Same Silver Moon". And he would waltz around the living room and the hall…. The older kids would waltz around with him, or they would circle Daddy, or be on the side and waltz around. It's come down five generations.*

That's one way Daddy soothed fussing babies. Years later I would discover the science behind his wisdom.

Table of Contents

A Note at the Beginning

Introduction

Part I: Cuddles and Cradle Songs

Recordings of the songs are available online at: https://bit.ly/LullabyWisdom

Part II: Simple Soothers

Endnotes for Introduction, Parts I and II

Glossary By Baby, science terms from the baby's point of view

Appendixes

Acknowledgments

References

Index

Recordings of the Simple Soothers are available online at:

https://bit.ly/LullabyWisdom

Dedicated to

Those who were my first soothers and sang me to sleep with serene
(and sometimes celebratory) voices
my mother, Laura Van Houten Bolton
my dad, Wilbur "Burzie" Bolton
my seven siblings,
Bill, Nancy, Helen, Betsy, Babsy, Molly & Robert
and also Rick who sang us into
love for a lifetime.

My toddlin' teachers: Jennie, Annie, Andrew, Joshua

...and all our children's children ...and their parents, all of the earth

"Lullaby, a human holy song"
– 8 year-old Laura Effa Van Houten

Introduction

"Necessity is the mother of mothering." -Dietrich Seaman

You've got a baby to calm, a toddler to keep happy. Help is in your hands! Yes, of course, I mean this caring book for all who want to soothe a child, a dementia resident, and *themselves* in the process. I also mean literally "in your hands" as soft embrace, and soothing touch. Perhaps you want to experience soothing. The stories, the pictures and the songs can float you there.

Enter One Grandmother

When my grandmother archetype greeted my psyche, she was a diminutive pepper-haired lady tending the warm hearth. The subdued coals glowed reddish as she stirred the suspended soup pot which steamed with the smell of onions, perking up my gastric juices to anticipate a shared supper. An image, of course! She left her tending, set the ladle on the ledge, and with delicate, agile steps, made her way toward me. Her smile let me know I was home, genuinely loved, and safe. Doesn't get better than that. *And* she came into my head for creating soothing rhymes and lines. Her gaze became mine: softened, timeless, available for as long as the infant or child keeps staring. Her presence softened my heart for soothing the babies and children by the hours, merged into my arms to embrace them for safety, my belly to prepare the foods of happy digestion, and my hands for sounds of play on the tummies. (The belly laughter is the reward, along with restful sleep after play.) She slowed my feet down to the patient pace of child exploration, sometimes only seven feet in 20 minutes to inspect fallen leaves or flowers, twigs or bugs. The complete grandmother! And where are the babies? Humanity needs tending!

"If there is anything better than being a grandmother, I don't know what it is!" – Joy R.

More Grandparents

The lullabyers in these pages embodied the grandmothers and the grandfathers with twinkles, playful interests, and loving quantities of patience. Their own lives were varied all across colors, cultures, sounds, and tastes of living. They loved the quiet, the family peace, soothing that spread into the evening. Yet the life experiences of some of these people came from events the likes of newspaper headlines and adventure paperback best sellers. Two spent 17 years in Egypt, and other years in Afghanistan and Sudan. Another was kidnapped by terrorists in Lebanon. He was the surviving hostage. Two had near death experiences when hit by cars. The indigenous grandmothers saw family lands swept from beneath them, and cultures denied by their children's educators. One lovely woman lost her mother, favorite soothing grandmother and then her father by 19 years of age....and was taken in by her best friend's mother. Several went through the difficult threshold of the death of a child.

Courage to Rebuild

Soothing isn't just for the easy, soft times, it is also for those who have been in the battlefield of life, and still have the sense and the compass, to build and rebuild families and caring communities. They nourish themselves, old ones, each other and infants.

Another family living in Southern United States, left their family and roots after a KKK death threat due before morning. In just hours, they were driving their family across many miles to somewhere safer. His crime? Running for public office. Still their hearts warmed each other and generations of young ones to come, as well as their trusted communities of church, work and home.

One traveled thousands of miles from her islands, yet loved the new people with her precious, serene voice and welcoming demeanor. Several soothers had mothers struck with mental illness who couldn't soothe them. Dads and relatives enfolded the offspring. Some had two alcoholic parents and did the tedious but rewarding work to flourish and open to themselves and others. Some had months of hospitalization. One provided legal aid to several unjustly-imprisoned, innocent residents at Guantanamo Bay Detention Center. One lost her husband shortly after her son was born.

Soothing isn't what you do because life is simply sweet, soothing gets you through the difficulties with kindness and tender touch strengthening the connections. The fires of deforestation sometimes fuel determination to dedicate the remaining life to growing the family tree. It's the subsequent delight and a shelter of care for those they love.

For those of us with more common lives, it is not that we need to go through hell to harmonize heaven. Instead, we desire to make everyday choices to serve life on earth and nurture the anxious and newly-arrived, then grow in joy. Building life can awaken all of us to the possibilities of peaceful living and shared laughter in a healthy culture of caring. World peace – two beings at a time. World peace – one being won.

Gathering Lullabies

Beside the grandparents mentioned previously, these songs and stories are also from parents, and people remembering grandparents. I additionally interviewed at the internationally-attended Haystack Rock Advanced Choral Conductors Workshop in Cannon Beach, Oregon where directors sing 200 new octavo songs in a multitude of languages and parts over five days. The conversation started with, "Did you sing to your kids? Did someone sing to you?"

They sang to me and very quickly I discovered the gold beyond the first vein of the song mine. With innocence-communicating faces they described scenes, their expressive hands touching imaginary infants as their voices gradually wove to velvet in various comforting styles. All were excited to pass on their joyful processes for stamina with infant care.

I collected songs over several years. Twelve out of the 32 songs in this book came spontaneously during the necessity of the moment: baby crying – do something. That something was singing and moving. The hope is that you, the reader, will embrace a way to slip a child toward slumber. Or slip yourself into winding down after a full day.

Lullaby Wisdom is the result of many hours of listening. In the editing, I endeavored to. stay faithful to each soother's words and their singing pitch. After 200+ interviews, I found that skilled singers are a technically different group than those intimate, liminal lullabiers who can bring on sleep in sometimes seconds. The groups overlap* but each group has its own skill set. The questions I've asked have foundational answers debuting as inner calmness, mindfulness, developmental science, and the positive feedback loop of an adult and child enjoying growing together.

*In **Singers Questionnaire** – Cannon Beach 2013, *Advanced Choral Conductors Workshop*, (L.Seaman), 65% noted that they enjoyed singing their whole life. Others picked it up as late as 25 years old, and even 40 years old. All these professional singers were singing six hours daily for five days, sight-singing 185 songs in arranged parts, with 125 other singers.

Cuddles for Brain Construction

The rapid developmental change in the world's newest infants is phenomenal. Humans grow between 80 billion and 100 billion nerve cells. Each neuron (nerve cell) has the possibility to form extensions multiple times with dendrites (branches) toward other neurons. Up to 10,000 dendritic connections (synapses) can be made by a single nerve cell.[1] The brain also has glia "support" cells, a trillion strong,[2] to assist the functions in the nerves.

Because the baby's head size is designed for holding a large, complex brain, the child is born before it is ready to be away from mother. Its head simply can't get any bigger before birth and get through the birth canal. Nature has also arranged the process so that even the womb time is a time of adaptation for mother's world. After the birth, the eager-to-be-held newborn continues the faster-than-imagination pace of adaptation. For our precious tiny one, the synapses that are forming, calculate into the trillions. Gabor Mate writes,

> *"The dynamic process by which 90% of the human brain's circuitry is wired after birth has been called 'neural Darwinism' because it involves the selection of those nerve cells (neurons), synapses, and circuits that will help the brain adapt to its particular environment..."*[3]

The growing and connecting is based on the experience of the infant.[4] Lack of stimulation, too much stimulation, and/or "bad" stimulation can make critical differences with such rapid formation. The first 2 ½ years truly is a time for developing a capacity for happiness, joy, purpose, exploration and a zest for life. Mate makes it clear:

> *"Positive **social** interactions actually build the PreFrontal Cortex (PFC) circuit by circuit until the circuits and myelination is fully mature. The PFC is the result of social experiences and is home to social understandings."*[5]

After this area behind baby's forehead is wired up by friendly, safe and responsive human interaction, the PreFrontal Cortex will account for developing internal emotional balance, empathy, compassion, kindness, and response flexibility. It helps us in reducing fear, gaining insight, forming a moral perspective of a larger good, building self-knowing awareness, as well as planning and problem-solving.[6] What a gift!

The Intimate Art of Child Development

Neuroscience is reminding us how amazing and life-giving it is to have human touch. Our tactile communications with each other have the possibility of optimizing and developing our brains' capabilities, and then increase our longevity and quality living, all while we are creating satisfying moments. Your hands, the way you caress your baby, share a world of information with the little one. We also contribute to our young children's and babies' dispositions, health, security, curiosity, braveness, kindness, attention span, problem-solving and brain power. Hopeful research cited within this book is uncovering these enticing findings about the importance of a baby having close, loving, and playful early childhood experiences.

Each baby is born with a powerful drive for intimacy. The baby <u>must</u> connect safely with another to grow all areas of the infant/toddler brain. Physiologically, Baby spontaneously and authentically connects with your voice. Because lullabies are essentially intimate, human experiences, they can convey ourselves, our worlds, our calm, and our sense of security, to the infant, smoothing the way for the baby's rapid, healthy development.

Sometimes soothing happens in the adult at the same time it happens to the anxious baby. That's the gift of oxytocin. Picking up the little one with loving attention can start oxytocin, dopamine and endorphin release in both people. Oxytocin is nature's solution to make sure the infant will survive; the cascading hormones bring intense rewards. In addition, sometimes soothing comes through sound.

Baby Bebel was still inside her mother's tummy. During the ultrasound for her 20 week check-up, she had fists that were very tight. The obstetrician wanted to observe fingers and thumbs on newly-forming Bebel, but she was too tense and wiggly to examine. Then her nine year-old sibling, Bijou, sang *Silent Night* over and over to her mom's belly, close to mom's skin. Later, the older sister related, "It was a soft and quiet song, and if I liked it, a little baby would like it." Fetal Bebs relaxed and opened her delicate hands. The astonished doctor remarked, "I've never seen anything like this....and....it is a girl."

Adding to touch, intimately-sounded melodies, like the one in the ultrasound room, embed deeply soothing memories. Many years earlier, when she was a toddler, the nine year-old in the previous story had experienced Mother snuggling with her each evening, softly singing *Silent Night* to drop her into her night-time sleep journey. In an intrinsic way, she could offer the comforting sweetness to her yet-to-be-born sister.

Stanford University's infant development researcher, Dr. Anne Fernald, speaks of sound's intimacies in an amazing way. Referring to mother-baby sound communication as 'Motherese', she said, "Sound can be described as 'touch at a distance'."[7]

These musical communications have their own brain receptors within us. Leading a team, Petr Janata used fMRI's to pinpoint the firing neurons. He found them in the rostromedial prefrontal cortex (rmPFC) behind the forehead. With precision, the rmPFC tracks musical pitches, melodies and familiar tunes as we hear them, yet changes subtly with each of our listening experiences, as it absorbs new perceptions and meanings about our musical memories.[8]

The infant in *#12 Baby's Singing Talks to Me* also had enough safety and intimate, personal music to develop her memory in the prefrontal area: she recognized the lullaby sung to her over the five months of her life, then spontaneously sang four phrases of it, on pitch, to her lullabying grandmother close to nap time.

Music elicits the richest autobiographical memories in us, working much like a film sound-track, but is much richer for us than either images or words.[9] The same area is also involved with directing our motion. Sounds like smooth coordination to me! As we intimately sing to babies or elders, we can cultivate soothing to last a lifetime. The informal and personal singing we share can truly add to our mental and emotional stability, and our happiness.

Why Do We Need This Book Now?

The purpose of this book is to encourage people to slow down for a moment, slip into inner calmness through mindfulness, and enter into the relaxed breath of the musical phrase. You can use these songs to embrace the young, and sing lullabies at nap time to enter into shared relaxation. These are family-building skills for a healthier society. More friendly connections can strengthen peace in us and our fragile world. The skills and melodies can also be used for adults with memory loss, dementia, or grief. This volume is for anyone who wishes to be soothed.

Healthy parenting may create many unforeseen positive effects. The recorded interviews were transcribed by a friend who had some adverse childhood events. Around eight months into typing the volumes of discs, she shared with delight,

> *"I feel like my childhood is getting healed by hearing these hushed, loving hours of adults honoring infants, and sharing kindness with helpless toddlers. Predictably, about three minutes into the questionnaire, their voices drop lower, slower and smoother in the story, and into a soothing world. Their soft memories heal me!"* -A.H.

Good parenting is healing to experience, even if the new experience happens after one's childhood.

In the last century, society has changed. Smaller families, nuclear families without grandparents or aunties, or families without relatives living next door, means that soothing ways may not get transferred from an elder to a youth. Lullaby skills are often passed generation to generation and may be lost with the increase of technological communication, use of personal electronic devices, distance, and busier lives.

One-child families may increase opportunities for the young one. Nevertheless, the smaller family unit may not allow a child to pass on what they learn of soothing to the next sibling. Earlier in the last century, many older children were mother's helpers, and took care of younger siblings. The eldest may have organically learned to soothe as early as four years of age, like the one great-grandmother in these pages (see #24), who has over 76 years of soothing one-by-one, brothers and sisters, her own family, and extended family infants: 27 in all! She and other ordinary people like her are "super soothers". Through routines at home, the gentle skills were passing down from mother to child, older cousin to infant, capable big brother to sick baby, or auntie to newborn, during the growing years. Now these pages hold the treasured expertise for growing happy, compassionate people.

We have a longer work week than the last mid-century. At the same time, we have more mothers working outside the home, more children in child-care and at earlier ages. Less parental contact, less nurturing touch and less intimate sound can mean more childhood stress.

The overuse of personal electronic devices may lead to babies and toddlers have less quality time with parents and adults. Psychologists have noted a new behavioral condition: jealousy toward the object of the parent's attention: the 'electronic sibling' rivalry. One grandparent observes,

> *"I think this book is very timely given the number of devices I see people looking at... instead of their babies and children."* -Frosti Talley

Your available contact time can benefit from *Lullaby Wisdom*. Sharing tips on quality touch, sound, vocal play, and loving stories, the book supports your health and the health of those in your care.

Too much stuff, too many choices, combined with too little time, has led to many experiencing symptoms of "cumulative stress" disorder. Our current speed of daily life and crammed schedules is stressful. Without enough free time, free play, down time, and relaxed social time with adults, a society creates stressed children. More structured and scheduled time is neither healthy nor the answer.

We know that enrichment is valuable, but how many of us are aware that too much enrichment (super-enrichment), is too stimulating for children, and produces stress?[10] Then, as a result of hyper-arousal, excessive adrenal hormones such as cortisol, inhibit childhood cerebral-cortical growth.

Babies crave people not products. Instead though, many parents increasingly find technological devices and canned sounds to endeavor to replace their human contact with baby or toddler during the parents' many busy times. This can limit infant right hemispheric opportunities for growing social intelligence during this 2½ year window of rapid acquisition and neuron-building. One mother remembers softly singing with her baby,

> *"I was really enjoying singing to Fiona as an infant when she was on my chest. I felt it gave her a very different kind of physical sensation from my body to hers. The vibration and all of that. I just thought it was very good for her neurologically to experience that. Oh, it was heavenly. I was very happy."* -Frosti Talley

This is nature's promise: we lovingly socialize in order to survive and pass life on. Ignoring this early developmental window of emotional growth can lead to a stressed society and a stressed planet.

These chapters and lullabies invite us to be closer to ourselves as a biological species in symbiotic relationship with our natural environment. Our very lives and the expression of our planet depend on humans attuning to each other and the natural sphere. We strive for intelligence and success, yet we are just at the beginning of the road of understanding our physical body-brain's interrelationship with climate, microbes, forests and the natural sounds, as well as the sights and the sensations that maximize the intelligence and success of our species. Cultivating human compassion with the most helpless, leads to greater world health, one child at a time.

Lullaby Wisdom holds explanations of scientific neurological processes that contribute to our health, and the health of the littlest. Some of the science within explains our tendency to resonate and thrive with nature. It is as if we and nature had grown up together as species and ecology. Indeed we have....through many eons!

The Neural Formation Of Kindness Takes Two

We are unusual as a species on the planet. We have developed four brains. We share our lowest brain's exquisitely functional design with all reptiles. It keeps our bodies running smoothly.

Above that is a brain we share with all mammals: the limbic or mid-brain. It processes emotions. Because our babies are helpless for such a long stretch, adult mammals have oxytocin and other neuropeptides of caring, to buffer them from the pain and stress associated with child rearing.[14] Oxtocin's partner, vasopressin, is part of our protecting circuit. In the middle of our larger brain, our limbic brain puts a tiny, emotional tag on every signal sent from cell to cell in our bodies.[15] Why? It does this to determine if we are safe, going to be safe, or are in threat from signals within our bodies, or from the environment, or nearby people. This reflexive process is called *neuroception*. It happens without requiring our conscious awareness.[16] We may notice an emotional change, but it is already in process. As mammals, our gift is social warmth and working collaboratively in groups.[17] Our safety is in our numbers.[18]

About two million years ago, our third brain, the human neo-cortex tripled in size for fancier thinking and nuanced communication.[19] One hundred thousand years ago, we greatly enlarged and increased the complexity of our 'fourth brain' behind the forehead called the Pre-Frontal Cortex or PFC. This area is not only unfinished at birth, it also is yet to be wired into nerve circuits with the rest of the brain. In fact, most of the neo-cortex and PFC circuits have yet to be built. Sounds hard? Not to worry, for we know internally how to frame the structures in three to five years. How?

Well, our needs are fairly simple, and yet not always available. **We need safety**.[20] Nerve pathways are laid in with safe, warm human interactions. Imbued with love's nourishment, the PFC grows rapidly. Through tactile stimulation[21] (i.e. hugs, snuggles, gentle touch, quieting strokes,) the process of branching and connecting neurons begins.

At birth, many of Baby's billions of neurons in the neocortex look like tree saplings. They are sprouted but need to fill out with branches, (dendrites) like a tree. Each of these millions of dendrites will grow to send messages to another cell by forming receptors and gaps called synapses. Then, electro-chemical signals can jump across from one cell to another. When Baby feels safe, they grow phenomenally quickly, setting capabilities for a whole lifetime.[22] [23]

Within safe, secure relationships, the power of another's love is remarkably seen in baby's rapid, optimum brain development. Baby's default nervous system is hunting for danger. We all have this automatic danger-detecting system for a lifetime. So the big challenge is to reduce threat, and/or feelings of danger. Being safe is more than someone telling us that we are safe, it is the feeling of safety from inside.

Then **kindness grows** within us. Even one touch experience of 15 minutes will increase dendritic budding, then change cerebral organization.[24] Our PFC not only gives us kindness, and compassion, but also control over our emotions, the ability to plan, to develop higher cognitive functions,[25] and to witness ourselves. These are the results of "presents of time" from caring people. Eye-to-eye contact, warm touches, singing, friendly sounds, and responsive conversation are the sensory richness laying the groundwork of PFC circuits, which blossom in adolescence.[26] The loving gifts turn off the Danger alert and put on the 'vagal brake', switching the Vagus nerve path from the Dorsal Vagus to the Ventral Vagus Social Engagement System. This system activates the heart, lungs, larynx, pharynx, middle ear muscles, facial muscles, head turning, and facial nerves to send social signals of safety.[27] Responding to, playing with, and soothing young ones will actually switch them to their safety circuit where their brains can grow exquisite nerve connections.[28] These well-loved babies will later begin to imitate those they trust.

Our brains have two sides, or hemispheres. The left hemisphere includes areas for language among its other capabilities. Much of the right hemisphere is devoted to reading the social cues of others: eye micro-expressions which can change within a second of time, facial expressions, voice intonations, gesture interpretations, posture: are these signals of safety...or of danger?[30] Baby also needs to grow more nerve fibers in the corpus callosum. which bridges the hemispheres of the brain.[29] Then Baby can integrate experiences, construct meaning, and develop language skills to interact with others. Loving is so simple, and yet nourishes the young (and all of us) for a lifetime.

A World of Soothing

I have collected global songs and stories from a Zimbabwean, a New Zealand Samoan, now in Alaska, a charismatic Brazilian lawyer who remembers a row of rocking chairs on the family's front porch overlooking the Amazon river for evening talk and rock, her son and an insightful nephew, an American second-generation hospital maternity nurse, even the conversational and radiant Irish-American woman behind me in the line at the post office, as well as my family (with trained singer heritage), my friends, my friends' international friends and families. Two grandpa 'baby whisperers' even recalled successfully calming fellow restaurant goers' infants. Yes, the pleased parents savored a relaxing meal.

You can enjoy stories of people from four continents and over 20 different cultures who share their "Best Practices": touch, holding styles, and voice volume changes from loud to whisper. Just humming the melodies invites your child into life as a global citizen. Add the languages, and your little one's neurons keep and develop a broader linguistic intelligence (as noted in #22). Languages for songs used within these pages are English (16), French (2), Spanish (2), Nahuatl, Samoan, Portuguese, Arabic, Sanskrit, Shona, Gallic, Native American, and non-word sounds called vocables. Using these songs and stories, you can encourage the awareness and respect for many cultures and peoples, and increase the possibility of creating peaceful, world communication. Soothed children develop their whole-brained compassion, calmness, initiative, and problem-solving. Their life-giving creativity blossoms.

Fig. 1 – Cultures and locations represented in lullaby interviews (partial list)

Is baby ready for many languages?

Yes, Yes and Yes! Babies' wondrous complex developing brains need live languages and close communication through a caring and familiar adult in order to remain active and avoid the neuronal pruning process of language acquisition that comes in early childhood. I chose serene songs in a variety of languages to inform the infant's brain for its maximum neuronal receptivity to language learning. The nap time routine is especially successful for strengthening language acquisition, for sleep reinforces the dendrites acquired with the person's most recent experiences,[11] and the lullaby experience before sleep is in a prime spot for excellent memory retention.

How to use this book – do one or any of these in any order, or all if you wish.

1. **Sing the song.** These selected songs sang *me*. They are that catchy! And yes, the songs are sleep-producing or calming.

2. **Gaze at a scene.** If you have a few seconds, you might just flip a page, and gaze into the peaceful picturesque watercolors.

3. **A serene story reads in a couple of minutes.** At the left of each song is the singer's tale recalling the richness of family times and the song. The stories sink us into a slower, restful stream. A glance here, a comfort there.

4. **Read the science following the icons.** Would you really like to know what is going on, driving the neurons inside your brain? How can you make the most of your time with your infant, toddler, child or elder? The **Baby Icon**, **Child Icon**, and/or **Adult Icon** on the story page indicate whose brain is changing. Can you both feel better, at the same time? Yes, sometimes both soother and soothee benefit. Perhaps you can apply the science to many areas of your life, then marvel at your wonderful neuronal adaptability that helps you to grow positive connections with others. Want more? In **Glossary by Baby**, these extra-smart babies tell you what is changing within their brains.

5. **Learn a new favorite song from the sheet music or the audio recordings.** The interview field recordings are examples for your easy learning. You can hear them sung at **https://bit.ly/LullabyWisdom**. Remember, your little one loves *your* voice!

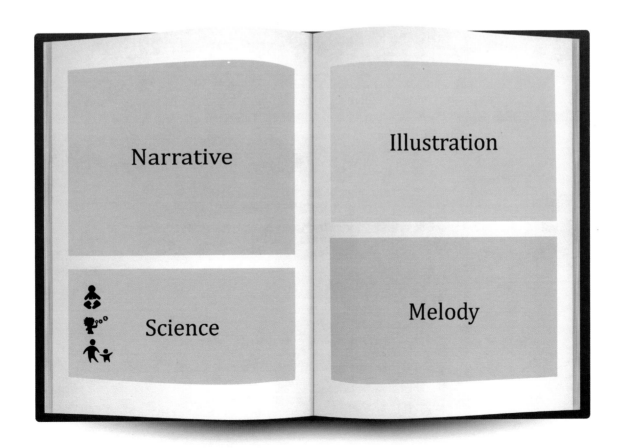

Lullaby Wisdom Part I: Cuddles & Cradle Songs shares 32 beautiful melodies to enjoy. The lyrics and musical staff are next to the story. Use one....or try a bunch. Part II: 'Simple Soothers' is for quick reference or times with the baby or toddler when words are too stimulating.

If you prefer, you can also learn from the online *Lullaby Wisdom* audio files found at this link: **https://bit.ly/LullabyWisdom**. Recorded at each interview, they are everyday lullabiers' spontaneous song renditions, made available here for educational purposes.

Over the years of gathering, I became aware of the subtly distinct, vocal qualities that a seasoned lullabier brings to the child. The sometimes intimate, drowsy sound is worth a thousand words. The contributors sing it better than I can describe it. The intent is for you to have enough sound to learn the tune, and the feeling, then sing in your own spontaneous way to create the maximum benefits for yourself and a baby, too. (See #22)

These songs are inherently drowsy in their tempo, rhythms and delivery. They soothe *me*, an adult!...and my sound engineer! They begged to be re-sung over and over in suspension of time. Simplicity is their gift. While the growing of inspiring children is a tender process, and soothing skills seem like a mystery, we can walk between the worlds of wake and sleep. Let's add more gentle voices to the Earth!

Many are the fancy recordings that lack the intimacy of a quiet voice. Though the commercial album may be arranged in an appealing style for the adult's ear, **the infant's or toddler's brain responds best to a warm, receptive, quieting human touch and a living, singing face with a voice that can grow softer and softer while listening to the infant's needs.** With a recording, the loudness cannot modulate to less stimulating, simpler hums and supportive whispers as the child grows sleepier. The skills and best practices in *Lullaby Wisdom* give us all (even the baby's uncle #21) a chance to be givers of peaceful moments until, responding to the baby, we become the 'baby whisperers'.

By soothing another person in song, you may grow the dendrites for your own social intelligence, while also encouraging a strong, healthy heart, flexible lungs, efficient muscles,[12, 13] and a more peaceful mind. Play around with what you're given; there is no right or wrong way. To a baby, you have the voice of an angel, the gift of heaven on earth.

Glossary By Baby holds some basic Interpersonal Neuro-Biology science terms, explained from the baby's point of view.

For Further Reading & DVDs holds music and basic neurobiological support information.

Interested in the **Interview Form** to collect your own memories and melodies? It is in the **Appendices,** as is the harp music arrangement for *'Julian's Lullaby'*.

Part I: Cuddles & Cradle Songs

1

One of us, a grandmother, shares: *She's getting born. Ella, my grand-baby, and very soon after, I'm in the room all night with her mom and dad who are sleeping. They're worn out and E's fussing and crying in the bed, in a hard bassinet they provide and it's very... I don't know, not the right idea of what to do with a newborn. So she was wrapped up inside and I just picked her up and held her in my arms and rocked with her, and that was it. I didn't want to wake the parents, so I didn't talk or anything, but she slept in my arms for hours until the sun came up. The Sunrise Song began to form then. Once we went back to their house, I stayed for the first two weeks. My daughter-in-law, amazingly, wanted me to be there.* **-KD**

The Scent Of Family – My grandma's smell was familiar and comforting to me. My mother genetically passed down the **scent** of her body. I and other new babies can recognize the scent of mother from among other mothers' scents. Because of scent cues, a family member or closer ones may have the easiest time soothing me.

Holding – Why does holding a newborn like me matter? Because when I reach out with a gaze, you look at me and hold me lovingly, I produce oxytocin, 'the cuddle hormone'. This soothes and protects[4] my nervous system, and allows my brain to grow optimally; emotionally and cognitively as well. Otherwise, below my consciousness, I am constantly on the alert for my safety. You *are* my safety; I don't yet know that I am separate. **"The brain is the most unfinished organ at birth. The brain grows to 90% of adult size, in the first three years."**[1] I will go from having merged with another, to emerging into me.

I don't require this holding and connecting with you at the same intensity for all my child life. In fact, what researchers found in the study of The Strange Situation[2], is that the 12-20 month-old babies like me that have a **Secure Attachment** with someone who has predictably and sensitively met our needs, these babies are the most calm, and self-soothed when mom leaves the room. Unlike insecurely-attached babies, they are friendly to others when they are with their mothers. And also unlike the other infants, these happy babies will explore their surroundings freely, curious to discover how things work. That's how I want to be.

One of my newborn cousins had to stay at the hospital three extra days without her mother. When she came home, she was fitful and anxious. Mother then held her all the time. When asked about the constant holding, she replied with a smile, "She needs it." Seven months later, Baby was smiling, enjoying people, and with her secure growing, she was sometimes scooting away from mom to explore around her colorful, enticing world.[3]

SUNRISE SONG

Sunrise Song

By Karen Deora

Sung by Kay Pasquesi

Hear this sung at https://bit.ly/LullabyWisdom

And it just depends on what comes to my mind at that particular time. It might be another tune and it might be other words. Most times there are other words. It's like moment by moment. Their interaction with you brings out different words to sing to them. So sometimes I just ask them, "Why are you so precious?" and other times I sing,

> *"I just thank God for you..*
> *And I just ask God to make you as His child,*
> *To keep you as His own because you're so precious to Grammy."*

It's the words that fit that occasion for that particular time. It looks like what it looked like when I was growing up. My parents sang to the grandkids. And many times when we were together, I would say, "You know your grandmother would be so happy just to see us doing this." If we were singing, or whatever we were doing, I will refer to them. And I'd say, "The reason I'm doing this today, I got it from your great-grandparents, from my grandparents, so it's just a tradition for us to sing and pray together. It's just custom, a family tradition, to do that." In the heart, and in my body ...yes... it's a good feeling all over." - **EH**

[I dream of a world where every baby hears and knows they are the sweetest little child in the whole wide world, for each love is unique and all love is One.]

SWEETEST LITTLE CHILD

I am a fine little baby. My healthy body and your love tell me that I am safe. My brain grows optimally, learning everything I can from you, your elders and what you enjoy on the earth. I am also learning all your concerns and what gains your attention. Your stories help me to grow quickly and smartly. You are modeling creativity processes for my baby brain to learn. That's what love looks like for me: freshness of song in the moment. I am learning to be in love with life! When you are **improvising a song** for me, sharing your happiness, the fMRI research suggests that you are globally activating most of your brain regions[1] You do have something peculiar happen: your brain activates its medial prefrontal cortex (mPFC), and appears to deactivate the dorsolateral PFC. [2] You seemed to be turning off an area of your brain that is called, "the right temporo-parietal junction, which is ...deactivated to inhibit distraction by irrelevant stimuli that might impair performance."[3] You sing in the present moment. Any dormant genes that I have inherited from my song-soothing family will now have the stimulus to unlock and express themselves like you do. It's called **epigenetics** (DNA's ability to produce thousands of variations to continually respond to living). I'll sing for joy, too. I belong in my family.

One study placed very **old mammals in with young** animals and toys. The elders showed a 10% *increase* in brain weight![4] This finding runs counter to other research showing that older brains have the tendency to shrink with years. So, consider this: the continually shifting, playful interchanges with young ones, may add cell rejuvenation and cognitive strength to my adult brain!

"Kids: they always give something back." - Jerry Lewis Manley

Sweetest Little Child

Edna Hicks
Used with permission

You're the Sweet-est Lit-tle Child in the whole wide world. You're the sweet-est lit-tle child to me. You're the sweet-est lit-tle child in the whole wide world, I think that's why I love you, yes, I do.

Hear this sung at **https://bit.ly/LullabyWisdom**

Five Generations of Lullabies "that dug deep into our roots and our bones..."

*This is kind of the earliest one and it's got the five generations going, which is kind of sweet. I heard it as a child. My Mom sang it some. I gave it to... my grandmother sang it to me. My grandmother is Elvina Maillet and she was French-Canadian ... Montreal, Quebec. That was the source of this. All the cousins in the family knew this one. I could probably find a cousin who knows more words than I do. I just make up syllables, because **do do** is 'sleep', and **fais do do** is to 'go to sleep'. And so there is only the 'do do' and the 'bebe' ['baby'] in this one.*

I would use the voice and the rocking. I would rub a hand, I would rub a temple. Mostly a temple because it's a sleeping aid. Just really lightly. [Use one or two fingers with a circular motion. This point in Chinese medicine is known as Taiyang, used to calm & relieve headaches. Try it for yourself, too.] *This was a nice shutting down for "M". And what I'm doing, is, I'm putting about three fingers in the middle of baby's forehead and kind of very, very gently, very, very slowly, running it down off the tip of the nose. That was a motion I used on her.*

Another thing I did was, just gently, not really put a hand on them, but gently rest my hand over the baby's head, a whole lot of cheek, a whole lot of the back of the head., almost the top of the head. And of course when they're babies, your hand covers the whole head, but to kind of cover up the ear, cover up any sound that was coming in, just the last of the shutting down, quieting down.

I always watched; children will respond to you. Some don't want their nose or their forehead touched. Some of them love the temple, some of them don't. You just find out what the child responds to and use that. -**MF**

 For my weary brain, the rocking "nah's" and "la's" in your singing bring on simple slumber; words are extra. Your sound tells my right hemisphere and my **limbic emotional neural pathways** that I am safe. These structures in my mid-brain process and express emotions. My sleep is secure. I love my life.

Your embrace, your soothing voice, your eye gaze and responsive touch are increasing my **oxytocin** levels.[1] Oxytocin triggers **endorphins**, then dopamine and more, and I feel ooo so good. I am learning to trust. *And* I grow optimally with healthier bones, muscles, brain, a better metabolism, and more immune protection against disease — all from oxytocin.[2] You are resonant with me when you watch for my responses and adjust to me. You are my Protector. Your song and touch nourish beneficial chemical circuits within me and you, and increase your happiness as well. We both get beneficial 'rest and digest' time to repair cells, and build our bodies. We're health buddies!

DO DO BEBE FRENCH CANADA

Do Do Bebe

French-Canadian lullabye
rendition by Marilyn Fleming

Sung an octave lower.

Doh doh Be - be   nuh nuh nuh nuh nuh   nuh nuh nuh   Doh Doh   Be - be

La   la   la   la   la   nah _____

" Do Do Bebe" can be replaced by other words, such as "Sleepy Baby..." and "Sleep my Baby."
La la's, Na nah's, Da-dah's, these can all be included or not, as you live into the lulling song.

Hear this sung at https://bit.ly/LullabyWisdom

I just remembered one more song. It was actually very special to me, [so much] that I still remember to this day. Mom used to sing to us. And I remember that it was very special because it has some gestures and stuff to it. I think you (to his Aunt Rosanila) might remember when I start singing it. It was actually one of the songs when nothing bad happens. Another classic. And it was another song that talks about calming it down, you know, "Don't worry, the stress of the day is over. Get ready to sleep." Just follow the melody, beautiful and very nice. And it works. Oh, I'd fall asleep right away. It was nice because just now it was funny, I was realizing as we were singing, that it is really calming me down....[I'm] not about to sleep, but it gave me a lot of peace, just now.

I remember my dad holding me, rocking me like this (a little bounce in the arms), leaning on my Dad's chest. and then Mom singing. Mom's doing the choreography and the music.

♪ ♫ ♪ ♫

When I was eighteen. I was in Canada . I lived with a family. She used to sing lullabies to her granddaughter, "K", like 'Hmm hmm', this lullaby with sounds and voices with lyrics. But not only that, what I found so interesting is, every time she fell asleep they used to put on classical songs that were just playing. It was really peaceful because the song, like the song on the radio when she was sleeping, was all through the house. So it was funny, because at the same time, you could see the baby falling asleep and having a peaceful dream, you also would feel peace and calm because you were listening to the same song that she sang while the baby was listening.

I wasn't sleeping. She was on the second floor. I was in the basement. But even while I was listening, I would feel calm. -**CF**

Mother's or Dad's touch on my eyelids begins the flow of calming neurotransmitters like **dopamine** and oxytocin within me. I am developing neuro-synapses predominately in the **right hemisphere** of my **neocortex** where my emotional cues reside, such as reading faces, other social body cues, and reading emotional changes in the eyes that happen in a fraction of a second, called **micro-expressions**. **Nonsense words** can be a calming and connecting technique involving the right brain. Human attachment expert, Pat Crittenden, says that strategies may use "nonsense words to convey unstated semantic meanings."[1] As my emotions calm, I gain neuro-synapses for life. Most likely, I will use them later as **implicit memory,** resilience, and more abundant access to my calm state.

Tum Tum Tum

Brazilian folksong
Sung by Carlos Feitoza, Antonio Lucas Feitoza
& Rosanila Feitoza Pantoja

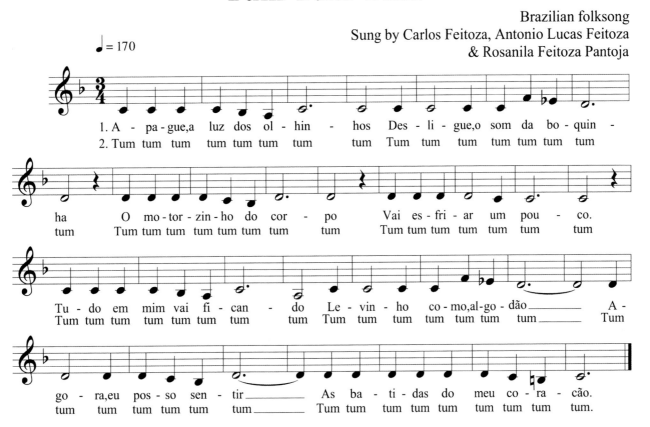

1. A - pa - gue,a luz dos ol - hin - hos Des - li - gue,o som da bo - quin -
2. Tum tum tum tum tum tum tum tum Tum tum tum tum tum tum tum

ha O mo - tor - zin - ho do cor - po Vai es - fri - ar um pou - co.
tum Tum tum tum tum tum tum tum tum Tum tum tum tum tum tum tum

Tu - do em mim vai fi - can - do Le - vin - ho co - mo,al - go - dão___ A -
Tum tum tum tum tum tum tum tum Tum tum tum tum tum tum tum___ Tum

go - ra,eu pos - so sen - tir___ As ba - ti - das do meu co - ra - cão.
tum tum tum tum tum tum___ Tum tum tum tum tum tum tum tum tum.

Translation:
Turn off the lights from your little eyes.
Turn off the sound from your little mouth.
The engine inside of your body is going to take a break.
It's going to rest a little bit.
Everything inside of me is getting so soft like cotton balls.
Light and soft.
Now I can feel the beat of my heart.   Tum tum tum tum....

Motions:
Fingers in a "V" tenderly touch baby eyelids, closing them.
Finger travels down the baby lips and closes them.
Fingers gently patter clock-wise around baby's chest.

Add light and soft touch to the fingers on the chest.

Hear this sung at **https://bit.ly/LullabyWisdom**

GRANDMOTHER'S LULLABY KENYA

I'm from Canada. This is a lullaby I remember my grandmother singing to me back in Nairobi, Kenya where I was born. And it goes like this. I don't know the words, but it goes like this: [song] And then it repeats. She would be holding me. --A.H.
[A.H.. lived in Nairobi, Kenya during the first six years of his life.]

I remember the melody and the comfort of Grandmother's arms. With oxytocin release caused by her embrace, my **amygdala**, part of my brain's alarm system, signaled safety. I modulated into my profoundly restorative Parasympathetic Nervous System (PNS). Grandmother's soft voice induced relaxation in me and slowed our brain waves from Beta (logic & reason) to Alpha (rest), allowing calmness to settle into me, and eventually, I dropped into the even slower brain waves of Delta and sleep.

SLEEP BRINGS A BRAIN BATH! Now I get my brain cleansing.[1] **Sleep** protects my brain from plaque by washing away debris. When I sleep, my glial cells shrink, my glymphatic "plumbing" opens, and waves of spinal fluid wash over my neurons, flushing out neurotoxic molecules that have built up during my waking. "We need sleep. It cleans up the brain."[2] **-Dr. Maiken Nedergaard**

When A.H. recalls Grandmother's voice and gentleness, his own voice carries the template of her style down the generations. His auditory nerve in childhood, not only picked up the soothing sound which calmed his Vagus nerve, but also picked up the very way to duplicate Grandmother's serenity.

Babies are born with healthy **mirror neurons**[3] which allows them to quickly imitate another person whose behaviors have meaning. The fluency of new imitation depends what their muscles already know.[4] In A.H.'s lullaby, the soft tranquility gently takes us to safety.

Grandmother's nurture releases **oxytocin** in them both, enabling them to manage stress, adapt to their environment and feel safe. It even optimizes their digestion, and benefits their gut probiotics called **microbiota**. With her embracing song, oxytocin increases resilience in them, improves their immune function, and protects their bodies with optimum biologic expression, facilitating their health and wellness.[5]

Grandmother's Lullaby

Rendition by A.H.

Hear this sung at https://bit.ly/LullabyWisdom

It was autumn...a sunrise of deep lavender pearl iridescence...clouds swiftly changing...within an hour, the day was downright sunny!

My first day to take care of the baby full-time was dawning.

*Shortly, her mother was returning to work. I thought, "I'm not going to get this sweet one soothed." Mama left. "Muffin" was three months, eleven days old. Later, I was gently bounce-walking with her and looking out the window reflecting, "How are we going to get through this day? It's only ten o'clock in the morning. How is this going to go? ... because she **loves** that mama." Then a song came through on that fall day. It went like this. "It's time for your mama to go now...."*

Whatever happened in that moment of singing, caused her to begin relaxing. The more I created, the calmer she got. It was like she got swept up in the part of my brain that was creating. She was experiencing it. Perhaps she was responding to my more relaxed state as I let go of anxiety and entertained creativity while embracing her with love.

*Researchers say that the babies are **hypnagogic**. That means that babies sponge up your responses to life. They know your attitudes toward anybody you meet or pass by on the street, in the moment, unconsciously, long before you put things into words or leave them unsaid. Or perhaps when I started creating songs, she calmed, wanting to know what the next rhyme was, or she was fascinated because I was fascinated. -LS*

 I have **hypnogogic responses**, meaning that I pick up your non-verbal attitudes and emotions, unconscious or conscious, easily through my infant heightened right-brain activity. I absorb your social cues.

"Well, if we want to think about a way of getting a taste of that kind of baby consciousness as adults, I think the best thing is to think about cases where we're put in a new situation that we've never been in before -- when we fall in love with someone new, or when we're in a new city for the first time. And what happens then is not that our consciousness contracts, it expands, so that those three days in Paris seem to be more full of consciousness and experience than all the months of being a walking, talking, faculty meeting-attending zombie back home. And by the way, that coffee, that wonderful coffee you've been drinking downstairs, actually mimics the effect of those baby neurotransmitters. So what's it like to be a baby? It's like being in love in Paris for the first time after you've had three double-espressos. (Laughter) That's a fantastic way to be, but it does tend to leave you waking up crying at three o'clock in the morning."

-Alison Gopnik PhD[1]

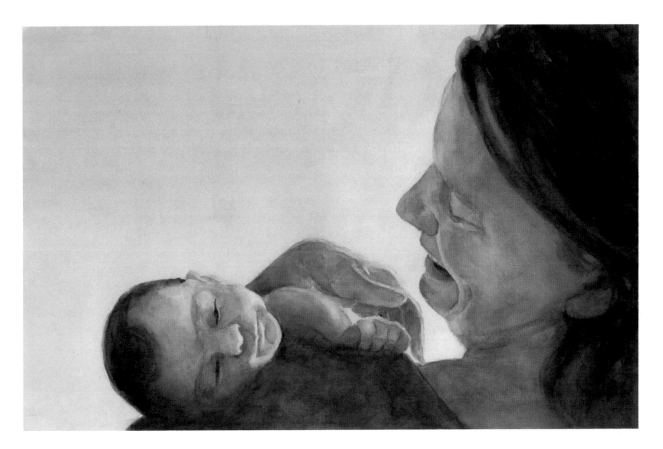

Your Mama Has to Be Gone Now

Licia Claire Seaman
3 & 4 by R. Seaman

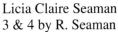

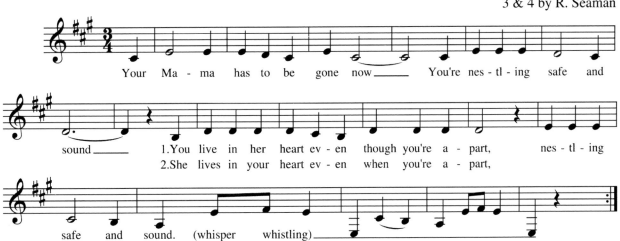

opt. verses: 3. Your mama and dada are gone now. Traveling far away.
You live in their hearts even though you're apart when they're traveling far away.

4. Your mama and dada will come back. Come back to see you again.
You live in their hearts even though you're apart. They're coming to see you again.

Hear this sung at https://bit.ly/LullabyWisdom

TU TU TESHKOTE 'AZTEC LULLABY' NAHUATL

*I*sing it and sing it and sing it. That's what I do.

My mother was the constant force that was there for us that could just, through embracing and rocking and singing, take every worry away and comfort us. My father was very loving and comforting. Humor and music were his gifts. The rocker was in my bedroom, I was the youngest. They were both very affectionate with hugging, snuggling, warmth and touch. They're both from the mid-West: my mother from Kansas, my father from Kentucky. I was raised in Alexandria, Virginia, and they, ...if someone came to the door, they would just give them a big hug and welcome them in. So it didn't matter if they'd never met before. That was home for me. Embrace. Touch. Warmth. Music. Laughter. Always love. Bedtime singing length just varied. I wouldn't say more than ten minutes. My mom had three children and had to get back to do the dishes and laundry and all that other stuff, so she was on the move, but she definitely was saying good-night sort of like, 'Did you have a good day? Was there anything upsetting you? We love you; everything's right in the universe. Have a peaceful slumber.'

Dad... he would sit next to me on the bed, probably with one arm around me, and stroke my face or my hair: stroke from the top of my head, back, the top of my head down, toward the neck, or the side of the head toward the neck. And always, my parents always looked me in my eyes. They always looked in my eyes. "Your world is complete. I am your world. The world is here for you. You are loved." Very loving. Other times, we sang for hours. Always.

*My dad's mother's rocking chair. It's my rock. I'm a rocker, for sure. In fact, the camp I worked at in Virginia was an old hotel with a porch and ten or twelve wooden rocking chairs. One class I taught with young teenagers was singing songs and rocking on the porch. -***KL**

My heartbeat loves *Tu Tu Teshkote* and I love you. Your singing happens to be traveling at half the pulse of my neonatal resting heartbeat. When I hear *Tu Tu,* my heartbeat feels supported and begins to synchronize harmoniously in with the beat of your voice. Your song encourages me to rest. You are attuning to me and I am becoming securely attached. You love me to happiness. I like this beat and I rest in it.

Hugs lower **cortisol** in my bloodstream and reduce my stress.[2] Hugs are good for you, too. We get fed in the cuddles. Hugging, we are a circle of life.

Humor sweetens us both. I learn it from you.[3] We are living happiness with our laughter.[4] Our shared friendly humor is much more about our social interaction and synchronized signals than about funny content. It is our social glue and greases our interactions.[5] With humor, my **immune system** is stronger, healthier, my **circulation** increases, my brain grows smarter, my **digestion** is easier, and I am enjoying life.[6]

You rock! **Rocking chairs** are ideal self-soothers. Even without the lullaby, the rocking chair can calm grown-ups and babies alike, helping many of us to feel safe. With the introduction of therapeutic rocking chairs at one psychiatric hospital, the patients who sometimes needed seclusion or restraint, were able to manage themselves 75% more of the time, through rocking in the rocking chair.[7] They changed their unpleasant emotional states from within themselves – it's called **self-regulation**. It's exercise, too! And exercise also can improve moods![8]

"We love you; everything's right in the universe: have a peaceful slumber."

Tu Tu Teshkote[1]

Traditional "Aztec" Nahuatl Lullabye
Version by Maestro Tlakaelel
Sung by Kathryn Langstaff

Tu tu tu tu Tu tu tu tu Tesh ko tay A - hah a - hah A - hah a - hah E - co - day.

Hear this sung at **https://bit.ly/LullabyWisdom**

8

At the end of the day, every day, before we went to bed, we were singing Samoan Polynesian hymns in our own language, songs for celebration and songs of the old culture. We'd gather in the front room, two grandmothers, Mom and Dad, the six children, and whoever was visiting from the extended family: fathers, brothers, cousins, uncles, aunts. At the end, they would leave if they weren't overnight. Whoever happens to be there. It's not so much, "Where do you live?" but "Where you staying at?"

Father would read from the Bible out loud in Samoan. We would be sitting side-by-side, next to each other. At first, Dad would start the first line of the music, then we'd join in a capella [without instruments]. Later, when we were older, he'd have each kid take turns doing the first line and others would join in. Adults would sit next to a child, and invite the child to sit next to the adult to help calm them. It was just expected. It was the ritual. At the end, we were quiet and calm. We sang in Samoan, the culture living in us, to keep it strong in the home.

We begin when they are little. Then as adolescents, they can come to their "identity, that's who I am, our culture, that's where I come from." They get into the messages that settle themselves. The family was all musical- the culture, the emotions. The family unit is traditional: the father, mother, brother, sister, all in the unity of coming together. This is very deep... to the core. It comforted. – **Missouri Smyth**

[Moe Moe Pepe is a Samoan nursery rhyme. M. remembers the lullaby. However, it was used at times other than the evening family group sing.]

MOE MOE PEPE SAMOA

They sing my culture. They sing my story. Within a few breaths, all our **group singers' hearts synchronize.** Swedish researchers found that when people sing together, synchronization occurs; heart-rates in the group accelerate and decelerate together. Our emotions sync up. Then our lung and heart rates change with the moods and speeds of the songs. With the slower exhalation used on most musical phrases, our singing creates beneficial conditions for cardiovascular function. The lungs and heart together called **Rhythmic Sinus Arhythmia** (RSA), and the form within a beat of the heart called **Heart Rate Variability** (HRV), are both positively affected. Our hearts beat as one.

In the above study, the lead researcher, Bjorn Vickhoff also notes,

"Singing regulates activity in the...**Vagus nerve** which is involved in our emotional life and our communication with others, and which, for example, affects our voice timbre. Songs with long phrases achieve the same effect as breathing exercises in yoga. In other words, through song we can exercise a certain control over mental states."[1]

Moe Moe Pepe

Samoan Lullabye
Rendition by Missouri Smyth

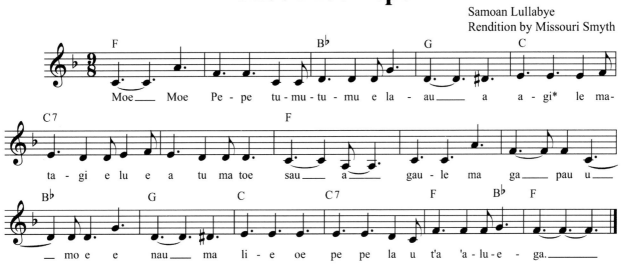

Moe___ Moe Pe - pe tu - mu - tu - mu e la - au___ a a - gi* le ma-

ta - gi e lu e a tu ma toe sau___ a gau - le ma ga___ pau u___

___ mo e e nau___ ma li - e oe pe pe la u t'a 'a - lu - e - ga.___

* In written Samoan, the "g's" are pronounced as "n"s.

You can substitute hums and "la la's" on this soothing melody.

Translation: *Sleep, sleep baby on the treetop*
When the wind blows, the cradle will rock.
When the bough breaks the cradle will fall.
Be pleased baby, because you rock.

Hear this sung at **https://bit.ly/LullabyWisdom**

9

I'd have them on my lap when they were little and then one on one side, and one on the other, when growing older. Both "M" and "A" really like being sung to. They both just calmed and settled, and I could get them to sleep pretty quickly. They liked their Daddy putting them to bed. - **PF**

His wife: *And if you're singing **'Go To Sleepy Little Baby'**. That was when our "M" and "A" were infants, which was way before that.*

P: *And my voice should be a little low anyway, to lull them to sleep.*

His wife: *You used to cuddle in that little head, right here by your chin and your neck, and hold their whole head in your hand and sing. It was sweet.* [Below the left cheek, and next to the heart.] *Hold that baby and sing. Put that baby back to sleep, please.*

P. sings: *Go to sleepy little baby, go to sleepy little baby.*
       When you wake, you'll patty patty cake,
       And ride a shiny little pony.

His wife: *You used it again with our grand-daughter when she was a babe in our house.*

As I cuddle skin-to-skin below his left cheek, and next to his heart, my head is feeling the touch of his tender hand. I receive a **"singing massage"**. I am resting my cheek on the warmth of Daddy's throat, in the vibrations of his vocal cords. My tiny body is rising and falling with the vibrations of the lungs, in their rhythm of inhalation and exhalation, while my heart beats its steady resting pulse with the rocking, rocking, rocking.

Daddy's vocal cords are also wrapped in a lattice of the responsive **Vagus nerve** which, all the time, is signaling safety to me: "You are safe in my hands." Sleep invitation becomes a peaceful pattern in my life. For my Vagus Nerve extends its pathway all the way through the spine to the viscera and informs my whole body. My body gets Daddy's vocal massage as his deep, harmonious vibrations stimulate my circulation.[1] In this moment, Dad is all I need.

My **auditory nerve** is directly connected to my Vagus nerve. So when you slip in soft, soothing sounds, you are directly calming my nervous system. When you lull with long, lyrical melodies, listening for my baby bodily changes, you are responding and adapting, changing your sound levels along with my sleepiness. This bi-directional loop of our signals in response to each other, you and me, me and you, contributes to my sensational R&R, in between my ever-ready duties of danger detection. I learn that the body is capable of safe, deep, deep rest. Above all, I gain the experience of pleasurable sleep.

"Hold that baby and sing."

Go To Sleepy Little Baby

American Folk Song
Rendition by Patrick Fleming

Swing eighths

Go to sleep-y lit-tle ba - by. Go to sleep-y lit-tle ba - by.

When you wake, you'll pat-ty pat-ty cake And ride a shin-y lit-tle po - ny.

Hear this sung at https://bit.ly/LullabyWisdom

10

Bismillah is Arabic. Usually I would end by humming. The words would drop out and then I would just be humming. Sometimes, they'd be in the crib and I'd rub their back. It could be just easier to have them fall asleep where they were going to be sleeping for the night instead of trying to transfer them once they got to sleep because sometimes that would wake them. So I went to stay there and rub their back and sing to them. Usually I'd just go in a circle, around, rub up the left side and down the other. That matched a yogic breathing pattern we were taught at one point, the Ida and the Pingala.

When we went camping and stuff, we would do that sometimes in the tent. I might have sung to them on the porch swing outside too. I think I did that once or twice. But usually as it was, I would sing lullabies as close to where they were sleeping as possible so that there was less trouble to move them, and it just made a psychological association between being sung to and going to sleep.

'Mr. Squirrelly'. He didn't just want me to rub the back, he wanted me to scratch his back. Just like you would rub, except using fingernails. He has tougher skin than our other kids and he just loved the scratching with fingernails in a <u>very</u> <u>light</u> <u>stroke</u>. Oh, once I'd start scratching his back, he would sometimes be out in less than a minute. It was amazing. I think it was sort of like a post-hypnotic suggestion that he would go to sleep once he started feeling that light touch. Yeah. Yeah. He really liked that. He wanted that all the way up until high school.

I feel gratitude because life is so precious and so fragile and it's just wonderful to be able to savor the best parts. I think putting little kids to sleep is one of the best parts of life. -**RS**

You rub me the right way. Mmmm. When you give me a consistent routine for sleep-time, I begin to make a **paired association** with the sequence of events, the song and hum, the back rub sensations and my onset of sleep. The secure and personalized routine soothes my default alarm system. This routine activity = sleep. In my brain, my **cerebellum** and **brain stem** get my body ready to slumber.

When I **hum**, my sinuses increase their release of beneficial nitric oxide (NO) by 15x. Humming contributes to my **sinus health** and increases my immunity. NO is naturally anti-viral, anti-microbial while supporting circulatory resilience in my nasal passages. Humming for at least ten seconds, five repeats, will better protect this beautiful child, and me, too, because it's healthy for upper-respiratory tracts.[1]

BISMILLAH ARABIC

"I feel gratitude because life is so precious and so fragile and it's just wonderful to be able to savor the best parts... putting little kids to sleep is one of the best parts of life."

Bismillah

traditional Arabic
Murshid Samuel Lewis, melody
Used with permission

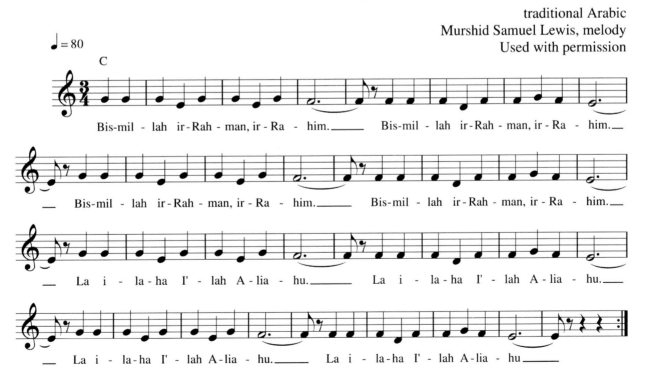

Bis-mil - lah ir-Rah - man, ir - Ra - him.____ Bis-mil - lah ir-Rah - man, ir - Ra - him.

Bis-mil - lah ir-Rah - man, ir - Ra - him.____ Bis-mil - lah ir-Rah - man, ir - Ra - him.__

La i - la-ha I' - lah A - lia - hu.____ La i - la-ha I' - lah A - lia - hu.__

La i - la-ha I' - lah A - lia - hu.____ La i - la-ha I' - lah A-lia - hu____

Translation: "In the name of Allah, most gracious, most merciful."

Hear this sung at **https://bit.ly/LullabyWisdom**

11

∽ SAFE IN THE NIGHT ∽

My grandmother, Mildred. Drake was her maiden name, Bradshaw was her married name. Mildred Bradshaw lived to 103. Well, that's the Wild West. Her mother was full-blood Cherokee. Her name was Catherine Rain and she married Willard Drake, who was Irish, redhead, fiery. But she left him because he threatened to kill his own child, "that little Indian girl." So one night when he was out, my great-grandmother fled with my grand-mother to Galena, Kansas and my great-grandmother set up a boarding home for the miners. She cooked for them and cleaned for them, and raised my grandmother. So my mother was named after Kate Drake, as was I. And I love being named after my mother.

Well, my mother is Cherokee and Irish, and her great-grandmother, Elizabeth Rain was on the Trail of Tears. This is a song that comes from Galena, Kansas and it's called **Rusty Old Halo.** *She would wake up if I was crying. I was right near her. My room was right off their room. And then she'd sing: "Shine like a star. Brighten the corner wherever you are." What was so wonderful about falling asleep in my mother's arms was the big rocking chair and the place on her chest that was home. I'm the youngest of three adopted children, and I had a hard time falling asleep sometimes, or waking up in the middle of the night. That place on my mother's chest and her words were definitely what rocked me back to sleep. She would probably hum, but it was the rocking. She was so good at just waking up in the middle of the night and just singing. I, having had my own child, I used to feel when she was nursing, or when she was here, ...**just being this world around her,** enveloping her. I'm sure that my mother taught me that.* – **KL**

RUSTY OLD HALO

That nightmare is no match for *my* mom! Mother held me to stop the nightmare DANGER. With our embrace, both of us began producing **oxytocin**, the **'cuddle' hormone.**[1] This feels good; it's a gift, the most abundant neuro-peptide in the brain's hypothalamus based on mRNA[2] Our brains are the primary producer of oxytocin, but other tissues, including the heart and the thymus (my mother's chest), can synthesize it at functionally significant levels.[3] With oxytocin, we trust each other. As oxytocin flows into my amygdala receptors (alarm), hippocampus (memory), lateral septum (coping strategies), and dorsal motor nucleus of the Vagus nerve, I now put on my 'vagal brake' which stops any FEAR. I'm switching to my Parasympathetic Nervous System. I can relax. I turn off my HPA axis (Hypothalamus/Pituitary/Adrenal) which started and fed fight/flight. Oxytocin is critical to helping me cope with stressors.[4] Brainstem nuclei that are oxytocin sensitive are part of the calming.[5] Either males or females with an oxytocin hug, handhold, or soothing sound can reduce my anxiety,[6] lower any inflammation[8] and strengthen our immune systems[9].

Oxytocin also increases **neurogenesis**, the growing of new nerve cells, in as little as 10-15 minutes.[7] And oxytocin speeds wound healing![10] Remember the elder with increased brain weight? Perhaps we *both* grow new cells. We're not aging....we're 'youth-ing'!!

A Rusty Old Halo

Written by Bob Merrill
Rendition by Kathryn Langstaff
As sung by her mother Cathryn Langstaff

rocking chair lilt 𝅗𝅥. = 46

Chorus

Rus - ty old ha - lo and skin - ny____ white cloud, Sec-ond hand wings full of patch - es____ Rus - ty old ha - lo and skin - ny white cloud, Robe so woo - ly it scratch - es.____

Verse 1

I know a man, rich as a king. Still, he won't give his neigh - bor a____ thing. One day he'll die. And you can bet, he'll get to hea - ven and here's what he'll get:

Verse 2

While you're on earth, shine like a star Bright-en the cor - ner where - ev - er you____ are. One day you'll die, and you can bet, you'll get to hea - ven but you'll nev - er get.

Hear this sung at **https://bit.ly/LullabyWisdom**

12

ᗯᗯ *BABY'S SINGING TALKS TO ME!* ᗯᗯ

We were using the Relaxing the Amygdala CD – 'Love ....' We had a forty-four minute version of it as a piano instrumental, and it just went over and over in a seamless melody. It had some pretty little things in there and some amorphous kind of sounds that lift you out of ... anyway, Rick, the audio engineer, put in the resonators. He got them tuned to the piano and then started this tamboura, this sitar kind of a sound. You know, that kind of floaty, cloudy stuff that happens in that Indian music. Well, he did that to the piano to form this beautiful, beautiful sound. We knew it was good and we took the CD to take care of our new grandbaby.

I also just sang it. Just, "Doo da doo doo." Just sang it to her. Five months. Keep in mind, I'm probably singing it four or five times a day.

Somewhere when 'Muffin" was around five months old, I heard her out of the blue, sing a 3-note phrase. She did it again. ..."Oo oo oo. Oo oo oo." And I thought, "Where have I heard that? Oh, it's the middle of the Relaxing piece because it goes....doo doo doo." So she had picked up the four repeating phrases, two slow and two moving. She sang that pattern, and I think she was trying to tell me she was ready for sleep. Because when she'd do that, from then on, after that, I'd say, "Fine, I'll put you down for a nap." And she'd drop off serenely and easily. She was using it for communication. That was the sound that came to her when she was desiring sleep. -**LCS**

My neural circuits are firing from my body through my **insula**, crossing over to the left hemisphere for language reception and production...and back down to my body in the process of being created. Repetition created the connection of the sound, rhythm, calming, expectation of my sleep state. A **myelinated circuit** emerges in me. At an early, delightful five months, I show the "wiring" – I sing the phrases. (Was it to soothe myself, signal you, or without any conscious intent – who knows?) You step in and complete the circuit by bringing it into relationship. 'Cause then *I* am telling *you* something about my state, and you respond with **contingent communication** (more circuit myelinating), and I do learn– I can signal you with that sound. So with this complex arc of having an **inner state,** noticing the state, labeling the state, sharing the state, having my state seen, having my state responded to in a contingent way [by Grandmother], my nervous system calms, my learning occurs: I associate expressing a need with safety/relationship and soothing... being a securely attached human is so simple and yet so complicated!

–analysis by Debra Pearce McCall PhD

LOVE, YOU ARE BORN

She sang that pattern, and I think she was trying to tell me she was ready for sleep.

Love, You Are Born
(round)

Licia Claire Seaman
lyrics for Relaxing CD

Love you are born. Love you are brea-thing us in.

My Love you are rai-sing the bar, in-vit-ing us to be-gin

All with Hope, Be-yond fear, Bring-ing Joy to the world that is here.

Optional Interlude

Duh-hun-duh nah Duh-hun-duh nah Duh-hun-duh nah Duh-hun-duh nah
(Tell a stor-y, What's a stor-y? You're a stor-y, We're a stor-y)

Hear this sung at **https://bit.ly/LullabyWisdom**

13

*My name is Christina Hubbard and this is a song I learned when I was singing in Veriditas Vocal Ensemble with Joan Szymko. It's just called **Lullaby** and it's billed as a traditional Swedish folk tune, or folk carol.*

I sang that to my son when he was little. -**CH**

I start the song with a moderate volume. Because I am singing, the in-breath is much shorter than the out-breath. This song has musical phrases that last for many seconds. My exhalation is often at least twice as long as my inhalation. When my exhalation lasts 7-10 seconds, such as with a singing phrase, my Vagus nerve signals my heart into a **Parasympathetic Nervous System** heartbeat, allowing my body-mind to relax.[1] Some scientists call this 'respiratory gating'[2]. The Vagus nerve puts its brake on my fight/flight response. My Vagus nerve is signaling to my brain's amygdala that I am out of danger. My heart gradually slows to steady rhythm. I am relaxing with you, and singing you into soothing, too. My body-brain has switched my Vagus nerve pathway to my newer **Ventral Vagal Circuit and Social Engagement System** that communicates from my heart, my facial muscles, and my eyes to you, and *your* facial muscles, heart, and eyes. With my tender gazing into your tiny face and the cuddles we share, we are building you into a lifetime of oxytocin-induced health benefits[5]. [See also #s 1, 3, 5, 11, 15, 28, 32]

I can further inspire you to enter into nourishing sleep by reducing my sound as I notice you getting lulled. I will sing softer, then softer, little by little, to provide less stimulation, more soothing, while keeping the rhythmic pulse like a reassuring heartbeat — the sound of your safety.

Wiring and firing my brain for happiness - Mom's or Dad's early closeness has given me safety to develop my **neo-cortex**, and build fancier nerve circuits in both hemispheres of my brain. I am secure enough to become interested in life around me. Besides the emotional and cognitive skills, I am also building the nerve pathways to my **Pre-frontal Cortex** or PFC (behind the forehead). This is not an automatic development in all infants; some won't have the face-time to fully connect their PFCs.[3] Forming the nerve passage-ways to the PFC requires loving, responsive and consistent care from you.[4]

Forming these nerve-circuits to the PFC in the first few years will allow me to self-soothe, be helpful, compassionate, and kind. This area of the brain allows me to develop patience, too. There's a lot going on inside me: nerves branching out, circuit building, new synapses firing — I think I need a nap!

Swedish Lullaby

Swedish Folk Song
Rendition by Christina Hubbard

Lie lie da low lah lie lie da low Lah lie lie da low lah lie lie da

low Lie lie da low lah lie lie da low lah lie lie da low dee

dee dle dee___ doh Dee did-dle-ly ah dun - doh Dee did-dle-ly ah dun - doh n

doh. Dee did-dle-ly ah dun - doh Dee did-dle-ly ah dun - doh n doh Lie lie da

low lah lie lie da low Lah lie lie da low lah lie lie da low Lie lie da

low lah lie lie da low lah lie lie da low Dee dee dle - dee___ doh.

Hear this sung at https://bit.ly/LullabyWisdom

14

I was raised in a large, caring family in San Francisco. There was always room at the dinner table for those who came looking for encouragement, companionship and conversation. Many years later, through Church, I met a 19 year-old political refugee from Cameroon. He had recently escaped his repressive regime when his life became endangered resulting from his human rights activities there. With the help of his coworkers and family, an airplane ticket was quickly arranged and he was taken to the airport and well-wished good-bye, opening a more hopeful alternative to what appeared to be his disappearing future. He left not knowing if or when he might ever see his family and friends again.

Fast forward ten years. He accepted an offer for a room in my home and completed a university education in pharmacy. He accepted a job offer for a major health research facility. He started paying back his student loans. The community has celebrated his marriage to a lovely woman of his ethnic background. He and his wife have told me that they will provide for me in my older years. The young couple purchased a house. Husband and wife lived for 14 months with me, which allowed them to build up savings. Dad is working. Mom is a student. Baby Zacharie is born.

One evening, "Z" was yowling and crying for his hunger to be relieved. His mother prepared to feed him. I was carrying newborn Z, cradling him in my arms with a soft, rhythmic bounce, when a song popped into me:

> *Mummy Mummy Mummy Mummy*
> *Tummy Tummy Tummy Tummy*
> *Yum _____ Yum _____ Yum*   **-JP**

MUMMY TUMMY YUM

Attachment: Hey Hey! I get it. I just need to relax...but how? It's you! The touch of your arms and breath, your voice and motion patterns, steady bounces, and repetitions are working their magic. I am soothed by you. Though you were single and had slipped past your child-bearing years without biological offspring, I recruited you by necessity and attached to you by care. In that moment, I saw Grandmother! Now waiting can be fun!

Inspiration: My calmness produced **Alpha waves** in my brain preceding a sudden burst of **gamma waves**. Gammas bring together information signals from all over the entire brain and I get an "Aha!" This inspiration happened in my right hemisphere of the neocortex, just above the ear, in the area called the **anterior Superior Temporal Gyrus**.[1] It activated just 300 milliseconds before I was aware. Suddenly the simple soother is there. I playfully try it out, lightly pulsing my arms. Soon after crying his danger signal of momentary hunger, Baby "Z" puts on his **Vagal brake**, and responds inter-personally with me. My motion and soft singing voice soothe his brain's limbic area, which begins calming his teary emotional state, allowing him to soak in pleasant sensations.[2]

"The touch of your arms and breath, your voice and motion patterns, steady bounces, and repetitions were working their magic."

Mummy, Tummy, Yum

Janet Purcell
Used with permission

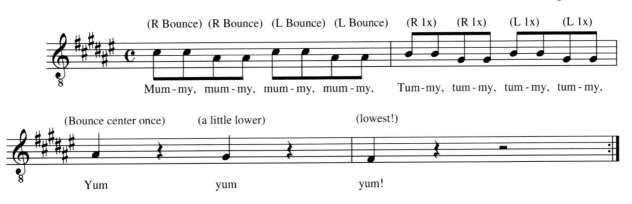

(R Bounce) (R Bounce) (L Bounce) (L Bounce)   (R 1x)   (R 1x)   (L 1x)   (L 1x)

Mum - my, mum - my, mum - my, mum - my,   Tum - my, tum - my, tum - my, tum - my,

(Bounce center once)   (a little lower)   (lowest!)

Yum   yum   yum!

As you hold the infant in your arms, bounce the arms to the right 2x,
Then swing the arms gently across the body, to the left 2x
Repeat: R 2x and L 2x
This time, move arms to the center to bounce just once "Yum".
Lower the arms just a bit, "yum".
And just a little bit lower, still close to your body.'yum!"

Hear this sung at **https://bit.ly/LullabyWisdom**

15

YES, JESUS LOVES "[CHILD'S NAME]"

My parents and lots of relatives sang to me. **It's a family tradition**...*so warm and inviting. Yes, and I can picture my mom sitting in this rocking chair, and there were times when she would have a bandanna around her head and the grandchild in her arms, and just singing to them. I can recall after a certain length of time, I would walk through the living room. Both of them would have fallen asleep in the chair. Then, when she would awaken, she would start back singing or humming one to the child.*

I remember my parents...nieces and nephews falling asleep in their arms, and asking them to sing it again. When they were older, I can recall them trying to sing the songs, and when they were falling asleep, half-mumbling the words.

In the summertime, we sat on the porch under the pecan trees, enjoying the fresh air...and played outside. There was a lot of singing. If the kids were crying, and had fallen and hurt themselves, we'd sing anything to soothe their little feelings and help them feel better, along with somebody's arm.

And when they were old enough to go to church with us, some of the songs that my parents did with them, they could sing in church. I thought that was really fine.

[retelling these memories:] *I just feel really good. It makes me recall so many memories that, otherwise, I wouldn't recall...so many at one time. But when you think about it, and you see this vivid picture in front of you, you can bring it to life in your mind and in your body. It's a good feeling.* **-EH**

Synchronization: Sing me into that loving feeling! Now *you've* got that lovin' feeling. When you sing my name, giving me the gift of a personalized song, I am learning that I am distinct from you. Experiencing your joyful voice, I hear that I am appreciated. The power of the music stirs your Grandma memories, too. I absorb your joy. I also have frequent social safety with my extended family members. In my family and during our congregational singing, we are all producing oxytocin which enables us to overcome our fears, create social bonds and love. I lower my stress, reduce my anxiety and elevate my endorphins.[1] **When we are singing together, our lungs and beat patterns of our hearts synchronize**∗; we are accelerating and decelerating our heartbeats in expressing emotions simultaneously as a group.[2] I learn that songs soothe.[3] Our soothing songs are for anytime, any 'owie', as well as for our family's and community's social and spiritual expression. Singing releases oxytocin which eases 'owie' pain and benefits wound healing.[4]

As rocking Grandma, I'm humming without being fully awake. I carry the soothing melodies deep in my **cerebellum**, subcortical regions and my respiratory patterns. With oxytocin's release, I soothe *myself* along with the baby. My half-mumbling child is carrying singing deep into the subconscious, too. The child in my arms, absorbs this from me through **mirror neurons** and loving embrace. Together we enjoy sleep!

∗Heart Rate Variability, (HRV) and Respiratory Sinus Arrhythmia (RSA)

Yes, Jesus Loves...

William Bradbury, 1862
Rendition by Edna Hicks

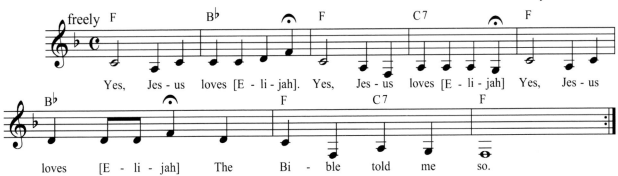

Yes, Jes - us loves [E - li - jah]. Yes, Jes - us loves [E - li - jah] Yes, Jes - us

loves [E - li - jah] The Bi - ble told me so.

Insert your child's name in here: [Elijah.]

Hear this sung at https://bit.ly/LullabyWisdom

16

*Whenever we spent the night at Gramma and Grampa's house, it was when we were there visiting. When Mom and Dad left, it was our private time with Gramma. Every night. At the end of the day, at bedtime, my grandmother would bring all of us into her living room and she would sit in this rocking chair. She did like us to sit with her. I remember sitting alongside her in the rocking chair and she would sing us this lullaby which is of German origin. I remember that, because she loved to sing it in German. I just remember the tune..dah, de dah, de dah, de dah [**Joseph Dearest, Joseph Mild**]...and I can just see her eyes, you know, just light up, and the joy she would have, it was almost like it was taking her back to when she was a little girl when her mother held her. We were fortunate to meet our great grandmother, although I never heard her sing, but I'm sure that music was really powerful in her life, because I look at my grandmother's eyes, I could see such joy, I could see her grandmother, and her grandmother's grandmother.*

Oh, it just seemed like forever, and I never wanted to leave because it was just such a comforting time. It transported me back to her roots. Even though she's Scottish, there's a little bit of German, but it was more just being back in time...for a while. It was probably wasn't even longer than the length of the song which she would sing over and over, over and over as she was rocking. You could just see her...you could see the story. That laugh, oh, goodness. You know, when you spoke with me the other day, I could see the memory that she was having, as a little girl being held by her mother. Even now, it brings back such joy. It just felt so comfortable there in Gramma's rocking chair. So, that's the story. You're conjuring up all these memories now. That's good! Where do I feel it in my body? You know I do feel it right in my gut, right now, you know? -**RB**

Eye-gazing – I receive an increase in **oxytocin**, the 'love hormone' when I gaze lovingly into your eyes.[1] Through **epigenetics**, the changing of DNA expression through experience, you carry that gift in behavior from those who loved you, your parents, and grandparents etc.[2]

Living joyful memories for grateful digestion – Right now I change from the remembering; I feel it in my **gut**. My abdominal intestines make 95% of my body's **serotonin**, a neurotransmitter for feeling good[3] and calming moods.[4] My gut also makes 50% of my body's **dopamine**(D2); the other 50% is made in my brain.[5] Dopamine is for learning, reinforcing, incentive in decision-making, desiring, and excitement.[6] My gut bacteria produce **GABA**, a mood stabilizer, which increases my serotonin. Besides these three neurotransmitters, my belly is home to 90% of my **immune system**. Perhaps the laughter and joy is stimulating my immune resiliency! The epithelial cellular layer of my intestines secretes **IgA** which prevents bacterial and viral infections in me.[7 & 8] My sensations are healthy memories: food for life! Take that, flu and colds season!

My Grandmother's Eyes

'Joseph Dearest'
German Traditional
Rendition - Rick Boyle

Da da dum da da da tum   Da da dum da dum da tum   Da da dum da

dum   da dum   da   dum   da dum   da   dah   da dum   dah   dum   dum.

Feel free to use a hum, la la's or any other sounds you relax with. It is less about
getting the syllables, and more about relaxing with the song, and enjoying the singing.

Hear this sung at https://bit.ly/LullabyWisdom

17

That was Gramma's song. Straight from Ireland. She was a pianist and just a darling lady. She was only five foot two. She came to Boston, and the only work she could get, because they were really down on the Irish immigrants, was as a maid. She worked really hard. Then one day, the lady she worked for, heard her playing the piano and came in and said, "Oh, what are you doing?" and my grandmother jumped a foot. She goes, "Oh, I'm sorry. I just couldn't keep from touching it." The lady happened to know people in the Boston orchestra, which back then wasn't the big Pops or Philharmonic or any of that, and suggested that she go there for a test to see if she was going to be any good or not, and she got the job, and that's how she met her husband, my grandfather. He was a violinist in the orchestra. Yeah, she got the job and she got the man. It was probably 1917, because Daddy was born in 1917, and she would have been in America when he was born. My grandfather was an amazing man.

Well, she'd sit on the side of my bed and hold my hand and first she'd read me a book, and then she'd sing the song, and then she'd give me a big kiss on the forehead if I was awake. I'm sure she did if I was asleep, but I don't know, because I was asleep. She was a very warm person. My nickname growing up and until this day, people call me "Dotty", but she called me "Dossy". For some reason that was her love name for me. She's the one that taught me to play piano. I'm not real good at it because we moved so far away when Mother got sick...to Los Angeles. I was ten and a half.

The lullaby: *When they [kids, grand-kids, great grand-children] were babies, I would hold them in my arms and sing. Then I would just kind of run my fingers across their eyes. Just two fingers. They're shaped like a "v" for victory and they're coming right down the eyes and the cheeks, all the way to the chin. [Then you'd do it again?] Right, because they had to close their eyes. I knew it was a beautiful song. I wanted them to hear it.*

They all know the history of my grandmother. Oh, I'd say pretty much what I told you, because I told them when they were older. I would talk about how great she was, how kind, loving, always hugging. Big hugger and so I'm a big hugger. **When I sang, I felt almost like she was with me. I just feel wonderful.** *I felt like it was helping to keep her memory alive. Oh, what a precious gift, yeah. Yep. The fact that she gave me the ability to be open and loving, and that's not always easy.*

Vocal skills: *You don't have to be fancy, you just have to care.*　　　　　-DL

Musical endorphins: When my newborn brain experiences lovely, gentle, rhythmic stimulation, it triggers relaxation. and the requirement of eye closure. Hey! It happens to adults, too. I put them to sleep as easily as they put me to sleep. Who need the rest here? Your music releases **endorphins** in me, as does your grooming touch, togetherness and love.[1] Over time, I gain from this song a direct cue to my full **implicit memory** of timeless love, security, hugs... and Gramma passed this gift along to me, in our relational experience. It's nourishing **emotional contagion** (the unconsciously-acquired emotion from another).[2]

"...first she'd read me a book, and then she'd sing the song, and then she'd give me a big kiss on the forehead if I was awake."

Loo La La Loo Loo

Ambroise Thomas
'Raymond' Overture
1851

The melody was later adapted in 1951 to the song "Hush-A-Bye" with lyrics by Sammy Fain & Jerry Seelen.

18

My grandkids? I revere them. They're precious. The oldest is forty-seven, then a younger generation [down to] nine. We made the song up and sang it to our kids; it became the family lullaby. With the grand-children, the song continued. And then it became, 'Mamma and Daddy and Baba and Nana and Thati love you very much' and it just kept growing and growing. **-JTS**

I do believe that when Natalie would come over sometimes in the afternoon and Jim would fall asleep, that Natalie would go to sing him the song. 'Bye Bye Baba.' Natalie is the one who is now nine. So when she was younger, probably three or four, she would come over to our house. She would sing to grandpa so he could have his afternoon nap. It is a family heirloom. **-MLS**

In my family, there are all kinds of family stories that are mostly true that pass down from one generation to another. My father was a country banker. So he loaned farmers money. In Kansas, in that period, it was easy to knock over a bank. They heard the explosion at the bank, because what they would do is bring in nitroglycerine and pour it all around the perimeter. And then they'd blow the door. And so my dad, his dad and his dad's brother headed out the door. Grabbed guns. We always had loaded guns, and grabbed guns. My dad who was the youngest grabbed a shotgun and was headed out the door. He was about five years old because he was going to join the men of the family, and his mother stopped him so fast that he didn't get out the door. We've told that story down the generations. **-JTS**

When grandchild Jesse was a baby...we were all going to have Thanksgiving dinner. Jesse had just been crying and crying and crying. So we were driving, and we were singing the song where he was in the baby seat in the back. It was just the two of us taking him over because Karen was already over helping her Mom. We sang the song and finally thought, "Well, he's asleep." We stopped singing just about the time we hit the Interstate bridge and it was very quiet. Then we heard this little voice in the back say, "More?" By all means we would, and did." **-JTS**

 Family for me! – When I hear the familiar names the adults are using day to day, I gain a sense of belonging to something bigger than just you and myself. You are putting my family's names in the song, and putting me securely in my family! I like it!

In this last story, I eventually experience safety. Their singing brings oxytocin relief. Then my nerves switch to the nourishing **Ventral Vagal Social Engagement System,** allowing me to rest and connect. I'll even sleep better.[1] I can now fill with curiosity. When the vagus nerve acts as a brake, stopping the fight/flight response, we babies are less distractible.[2] We slip into focused attention and rewarding social interactions.[3] We are relaxed and interested in our environment. All together, we are nurturing an **inter-generational belonging,** for you are patient with the weakest among you.

 I am completing the cycle of CARE. I am modeling the caregiver. As a four-year-old, I am singing the soothing lullaby for my grandpa "Baba" to nap.

ROCK A BABY BYE

Rock A Baby Bye
(An 'adding names' song)

James T. Stewart
#2 sung by Mary Lou Stewart
Used with permission

Bye bye— ba - by, Bye bye— ba - by girl. Bye bye— ba - by mine

rock a ba - by bye.— Now it's time to go to sleep. Lay your head right down.—

Mom-my and Dad-dy* love you.— rock a ba - by bye.—

* Add as many names as you want.

An example using the 2nd version for baby Natalie:

Ma-ma and Dad-dy and Ba-ba and Na-na and Tha-ti and Jes-se and Ma-ia and I - an

love you._____ rock a ba - by bye._____

Hear this sung at **https://bit.ly/LullabyWisdom**

19

Sometimes, walking with the baby in my arms is quieting for me and the young one. I'm in good shape and can keep up the walk for as long as Baby needs, for going to sleep. As the baby gets older and heftier, I move the toddler to the over-the-shoulder position. The toddler's warm cheek rests on my shoulder. I consider it upper-body, strength-based exercise, and the walking stimulates metabolic processes and supports digestive health, as well as lowering heartburn tendencies. The steady beat of this song was born from the Dances of Universal Peace. Indeed, in the seventies, we danced in the group concentric circles at The Old Church to learn them, and indeed, I felt more peaceful. Often I danced with a baby on my back , and later, a child at my side as well. **-Mama LS**

Well, I guess I've sung in Sanskrit, too: **Sri Ram Jai Ram.** *We used to dance around, with her in my arms and I would sing that to her when she was just tiny. It felt very calm and restful, very warm and sweet. You know, it's hard sometimes to keep yourself from falling asleep when you sing lots of lullabies. You have to wake up again once you've got the little one down. Sometimes it's not worth it. Sometimes you just go to bed yourself; but sometimes you have things to do and you walk around, have some tea, whatever and get on with the rest of your evening.* **-Papa RS**

SRI RAM, JAI RAM SANSKRIT

Mindful Singing – Singing easily can be a mindful activity, allowing the images of the mind to stay with the song. I visualize sleepy images with minimal energy. I allow myself to hold this slower state of mind without falling asleep. If I am standing, sitting or rocking, I am mindful of keeping a gentle, secure hold on the infant or toddler. Then I see myself sliding in between the worlds of sleep and wakefulness. It is as though I sidle right up to the edge of sleep, standing next to the world of dreams, then deliberately keep an observer part of myself watching me, as I sing the lullaby. I let my mind relax and let the song sing itself, riding slow at sleep's entrance, without going into slumber. I choose to steer clear of energizing thoughts. Where the attention goes, the energy flows. If a thought comes, I neither attach to it, nor push it away. Instead, I softly watch it go by. I return my undivided focus back to the slumber descent of the little one who is also detaching from the world of doing, then entering into the realm of being.[1] Restful sound and repetitious gentle waves are all that is.

It's hard
sometimes
to keep
yourself
from falling
asleep
when you
sing lots of
lullabies

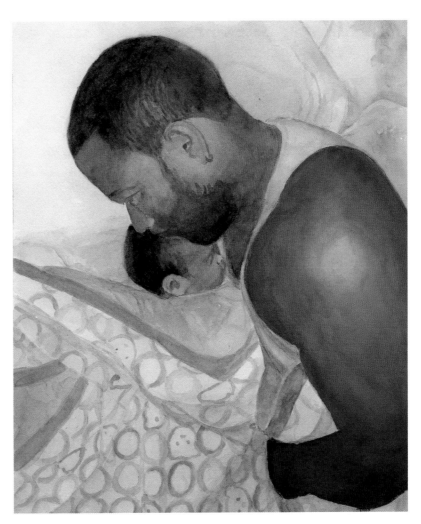

Sri Ram Jai Ram

words traditional Sanskrit
melody by Mushid Lewis
Used with permission

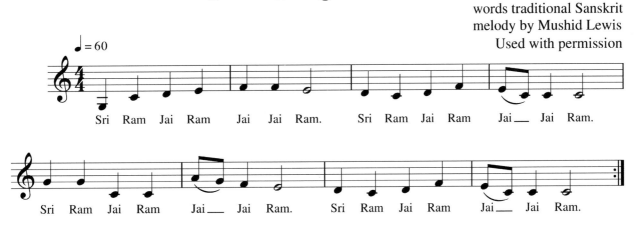

The words express devotion to God. Using them purifies the heart.
Sri: Resplendent, glorious, revered, respected;
Jai: Victory;
Ram: Spirit/Lord as light, strength, and virtue.

Hear this sung at **https://bit.ly/LullabyWisdom**

When the babes... especially when it was a tender day and the sleep was a little less solid and maybe they were not so happy, and so healthy that day; I would use the voice and the rocking and everything I could to cuddle them. At the end of lots of songs, I would be down to a whisper and it's hard to sing without doing the vibration, without vocalization, and just keep it in the air, and so I would end up with the, "Na na na na na na na na na ...al-le-pun" whispering.

...and you lose your tone and your melody. You actually lose it all, and so all you end up giving the children is that sort of wonderful "ruah" breath. [Hebrew for 'spirit']   I just give them the breath toward the end of a lot of the songs and watch them sink and sink and sink and it was very dear to be able to give that.

When you get to those quiet places and you thin a song and the baby's asleep and you want to capture that, you don't want to stop and take a breath. That's an interruption. You don't want to interrupt this beautiful thing that is happening. So you sing in and out and you breathe in and out and you go, "la la la ..." and it's funny and you kind of laugh inside, but it's fun. -**MF**

My **laughter** inside keeps me healthy; it relaxes me. Now I am more available to the child.

Liminality: Singing gradually softer and softer internalizes my focused attention until I, as singer, begin to embody the slower delta brain waves that lead to sleep. For me, the trick is to ride the edge between sleep and waking, so that the baby is securely cuddled, then modulate my consciousness to come back out without adding my attention to such thoughts as, 'To Do' lists, or chores. Liminality means 'threshold'.

Observing my mind: Around the sleeping infant, I become the observer of my mind, keeping my internal thoughts quiet through noticing them going by, outside the window of my mind, letting them be, until I am out of proximity of the child. I allow the drifting one to settle into deep sleep, undisturbed by my physical body's electro-magnetic field and mental chatter.

I continue the softness with my own thoughts and movements as I withdraw from the infant's location. If necessary, after a little one has a fussy waking period, I withdraw from the infant gradually, in increments, relaxing and softening my mind with each change of my slow movements.

ALEPUN SPAIN

Alepun

Spanish Villancicos
Rendition by Marilyn Fleming

Nah na na na na na na na na na A-le - pun   Nah na na na na na na na na na A-le - pun.

Du du du du du du nu nu nu A-le - pun, ya-le-pun, ya-le-pun, ya-le - pun, ya-le-pun, ya-le-pun ca-ta-

pun ___ Na na na na na na na na na na na na na na na na na na ca - ta - pun ___

La Virgen va caminando alepun (2x)
Y va por esa montana.

De quien es esa gallina? alepun (2x)
Y'el gallo qui est en corral?

La gallina est del cura alepun (2x)
Y'el gallo del sacristan.

Translation: *The virgin is walking*
Up the mountain.

And who has the hen?
And the rooster?

The hen is the priest's hen.
And the rooster is the sacristan's.

Hear this sung at **https://bit.ly/LullabyWisdom**

21

I am the grandson of Will E. Risk, a mule-skinner. He had a general store up in Nome, Alaska with his sister and his brother. They were true Wild-West people. He came in after the Gold Rush up in Alaska. My grand-mother, Cora Smith was a Victorian impersonator, one of the few MA Masters Level educator-women during that era. She went from the mid-West to teach Indians in North Dakota. Somewhere along the line, Will traveled up. He was in contact by letter, came up from Parsons, Kansas, away from mule-skinning, picked her up on a train, and took her out to Washington State, Napel region, now called Moses-Coulee. She taught the kids of miners up building the tunnels for the cross-state Cascade Railway. Anyway, he got his land, co-homesteaded with Cora. And he was a big thinker. He planted cherry. A cherry orchard takes some kind of vision, hope, planning and courage because it's a long time from planting to harvest. In the process of doing that, he and other homesteaders... what we would call a think-tank, 'troublemakers', instigated what turned out to be called Grand Coulee Dam. He had a vision of "...if only they had water." One thing lead to the other, and of course, that's now history.

I was born in Winatchee, in 1944, at the end of World War II. For the first four and a half years of my life, the grandparents lived right across the street. She would put us to sleep in the big bed. Then we'd be carried over to our house when the parents came home from work. Or ...there was this big rocker... with big arms, just a classic big rocker that all three of us could fit in with "Papa Will".

Frere Jacques: *Grandmother... she sang it with a punch during the day. You could feel the rhythm within her. At night, it was more a swinging lullaby, more soothing, as our rowdiness prevailed. (laughter)* **-Raymond W. Peterson**

The Singer: *When I recorded Frere Jacques, it started off louder than when I finished it. It's descending and pulling me in... a lulling effect, more than a nursery rhyme. So... Brother John hit the snooze. Why did I use French? Those are phonemes she wouldn't hear. Neurobiologically, there's a chemical reaction in the brain.* ***The sooner you are exposed to languages as a child, the more aptitude you have for acquiring or learning a new language.*** *-* Linguist **Joshua G. Seaman** for his CD, *Comptines et Berceuses por Bijou.* [See explanation on the next song - #22.]

As we share positive emotions[1] at bedtime, **laughter** is my gift to all, supporting wellness in me, too. My **play** forms healthy social connections with you. 20% of "what's funny" is content; the rest, 80%, is our social interaction, our playing off of each other, and our charisma.[2] Giggles and exercise oxygenate and safely relax my body-brain before I go to sleep. Children laugh more: we are happier.[3] We delight in being alive. That is our present to you. You adults might join us in the silly moment. Laughter increases our creativity, and improves our problem-solving ability.[4] Hilarity reduces pain and stress, then increases our endorphins and dopamine.[5] Mirth especially, helps me to grow in self-esteem and merge into my group identity of family and friends.[6]

Frere Jacques
(Are You Sleeping?)

French Folksong
Rendition by Joshua G. Seaman

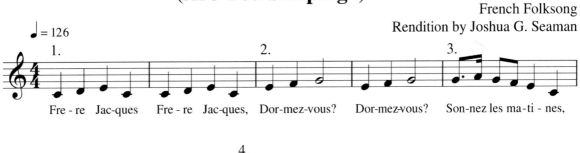

Fre - re Jac-ques Fre - re Jac-ques, Dor-mez-vous? Dor-mez-vous? Son-nez les ma-ti - nes,

Son - nez les ma - ti - nes: Din Din Don, Din Din Don.

translation.:

Are you sleeping? Are you sleeping?
Brother John, Brother John?
Morning bells are ringing. Morning bells are ringing.
Ding Ding Dong,   Ding Ding Dong.

Hear this sung at **https://bit.ly/LullabyWisdom**

22

~⌒~ 'SVUSVURUDUZA' ~⌒~

We take that baby onto our back and tie in a wrapper. As we are making this movement, the forwards and backwards from the waist, we are trying to make the baby sleep. Then we sing for the baby. Yes, we can sing over and over until the baby is asleep. We keep singing for them again until they sleep, then we stop singing. Now it depends with the age of the babies. For example, zero to six months when they are about to sleep, those babies, they cry. You know their cry is almost endless and it's very noisy, then you can detect that "Oh, this baby, wants to sleep." In fact, before you detect that that baby wants to sleep, you try breastfeeding, then of course, they keep on crying. So what we do, we take that baby onto our back and tie in a wrapper and, you know, we make these movements. The baby begins to be quiet. Slowly, slowly and begins to sleep. We make sure that the baby is asleep. Right? We nicely take the baby off from our back and lay down the baby on the bed. Or if you do not have a bed, you can just spread out your mat and some few blankets down, take a pillow, a small one for a baby, put that on the blankets, and you lay that baby on the side. Then when the baby is asleep, you tap lightly just a little bit...

- Patience Chaitezvi Munjeri, mbira ceremonial musician and author

[Svusvuruduza is the name of mother's whole body in motion, as she rocks the baby to sleep. It is done sitting in a chair or on the ground. When Patience was rocking in a chair, she was not only using her waist as the hinge, for the body; but she was also lifting the weight off her feet when she went back, and then coming forward. Her whole body has the rocker-framed motion, carrying that impulse back and forth. Her feet would lift slightly from the ground, keeping contact but no pressure, as a part of all the rocking system. Patience's whole hand is giving a *very light* touch. Mostly, the whole fingers are patting in a rhythmic pulse, 86 bpm.]

When I sing, my muscles work more efficiently.[1] **Muscles uses less oxygen during exercise while I am singing.** I am exerting less, singing more, as I rock. My muscles are also making **PGC-alpha-1,** saving my brain from depression.[2] My muscles also benefit from the oxytocin that singing[6] and exercise[7] both produce in me, making me more disease-resistent.[8]

The Language-Learning Window: My infant brain has many neurons specifically designated to hear the diverse sounds, inflections and articulations of all the world's languages. I'll put into use the living sounds and words that I hear from a person that I trust.[3] Your singing and talking are building my brain. "...The critical period for phonetic learning occurs prior to the end of my first year..."[4] By the age of three years, my brain will enter into a pruning process to discard the ones that have not yet been stimulated;[5] that is, I haven't heard *you* speaking or singing them. So sing and talk me into early language acquisition. It is just easier, earlier. Whoa...bring them on!

...we take that baby onto our back and tie in a wrapper and you know, we make these movements. The baby begins to be quiet.

Shona Lullaby
(Oh my little baby stop crying)

trad. Shona, Zimbabwe
Rendered by Patience Chaitezvi Munjeri

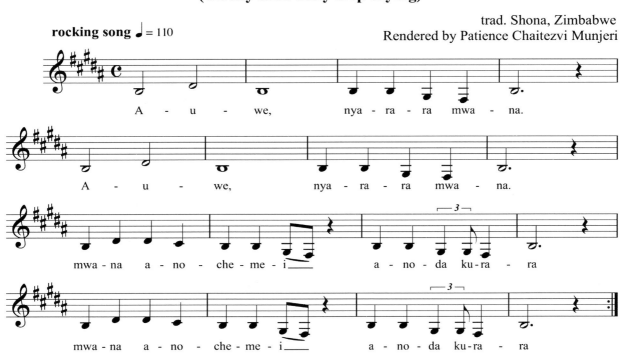

translation: Oh my little baby stop crying (2x)
What is it that the baby wants?
Oh s/he wants sleep. (2x)

Hear this sung at https://bit.ly/LullabyWisdom

23

*L*ast Thursday, Grandfather "Poppy" stepped into Grandmother's shoes for afternoon sitting. Poppy picked up three-year old "J" after his preschool.

Poppy: *Well we had lunch and then he wanted his milk. He likes it warmed. It's in a little sports bottle. He likes to be rocked when he drinks it and he likes to take off his shoes and socks. We went up to the room next to where he was going to nap and...*

Grandmother: Did he tell you we use the upstairs rocker? Did he say?

Poppy: *Yes, he wanted to go to the upstairs rocker. Anyway, he said it was in here. So we went in the sun room. He showed me where the massage oil was for rubbing his legs and his feet while I was rocking him. Then after he was done drinking milk, I rocked him. I started to sing to him. I started singing* **Bismillah** *(sung in Arabic).*

J: *(quietly) "No, ...don't do that song. I don't want that song."*

Poppy: *"How about* **The Green Worm Sleeps in Silk?"**

J: *"No! (sweetly) Bijou likes that one. (gently) No, I don't want that song."*

Poppy: *"How about* **Twinkle Twinkle Little Star***?"*

J: *"No,(sleepily) I don't want that song."*

Poppy: *"Well... (conversationally) what song do you want?"*

J: *"**Heya Heya Noah***?"*

Poppy: *"Oh, okay."*

So I sang **Heya Heya Noah** *and then in about a minute or two he was asleep and then I sang for about a minute or two after that, so that his breathing was deep and relaxed. Then I carried him off to bed. He slept a long time. He just felt very relaxed and safe. He is still sleeping.* **-RS**

My happy three-year-old brain, has internalized the consistent nap routine. I communicate to get my needs met. I tell Grandfather which rocker to use for naps, where to find massage oil, and which song takes me into sleep. I am **self-regulating** through my own self-knowledge of my internal states. I have a strong, secure relationship with both grandparents and I am able to transfer the routine from one to the other. Unknown to Grandfather Poppy, Grandmother uses only one song for my naps now. I trust Poppy to help me find it, and satisfy my mind's **invariant representation**[1] (model) for sleep preparation. Invariant representation is determined by a physical structure in my neocortex that recognizes models in my memory for animals (i.e. what a dog looks like), plants, objects, routines, etc. This structure in my neocortex is columnar and 6 cells deep with bi-directional signals which perceive the new information, then send it down to the memory pathways to check what I already know, then back up to the neocortex with the information: 'looks like' / 'doesn't look like' a nap routine. It takes me just milliseconds. Grandfather patiently reciprocates and strengthens my grandson trust. I get my wish: sleep.

Heya Heya Noah

U.S. Indigenous Spirit Singer, gifted to his grandson
Used with permission
Sung by Poppy & Amah

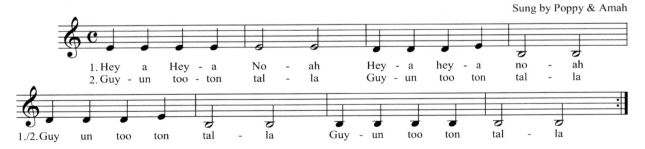

1. Hey - a Hey - a No - ah Hey - a hey - a no - ah
2. Guy - un too - ton tal - la Guy - un too ton tal - la

1./2. Guy un too ton tal - la Guy - un too ton tal - la

The lyrics of any Spirit Singing are beyond translation. They are essence.
As a lullabye, the sounds are the essence of sleepy.

Hear this sung at **https://bit.ly/LullabyWisdom**

24

Well, I've got nine grandchildren, and seven great-grandchildren, and I've taken care of all of them at one time or another. ... and now I have the little ones. When I'm cuddling the little ones, I automatically start doing it. ...and I'd just repeat that over and over again. "Duermete mi hijito" is "Go to sleep my little son, close your eyes." Towards the end where I could see that their eyes are starting to close, then I will lower my voice even more and just make it fade away and watch to see the reaction and see if they are going to sleep, or if they are asleep. They'll go to sleep every time. Probably in one to five minutes, because I can always tell when they're ready to go to bed. You can tell when they're getting tired, by their body language, how tired they look, how their eyes keep closing on them. What else? They just slowing down enough where you can tell that they're ready for a nap.

Well, I don't sing at all, and I definitely do that. It works for me and as far as I'm concerned, that's all that counts when I'm trying to put these kids to bed. Yeah, that over and over again. I know I'm winning when they're fighting and closing their eyes and fighting and closing. And they're fighting it and they usually lose the battle. They're tired. It's time. You know, when you're trying to put these children to sleep you will put any combination whatever, [words and syllables i.e. 'da, da da] *that might work. I don't know how I started or when I started, but once I figured it worked, then I used it.*

When it's time for nap time, I sit in the rocking chair. I rock them and I hold them and I pet their head around their face, like this. [She's showing me a very light, slow touch.] *Above the eyebrows and around their little cheek. Around the cheekbones down to the chin. And when I do that they stop... Always. And I'd pat their hair and my hand through their hair, just smoothing it down. I've done it to all of them.* — **-DF** - 'the baby whisperer'

L: You must be proud of your granddaughter.

DF: *Beyond that.*

Grandma Dee's singing activates her brain hemispheres to fire together as neuron signals pass across her brain's central fibers: the **corpus callosum.**

I respond to Grandma Dee's melodic rhythm which is inviting my brain to synchronize to the alpha speed in her notes, 7-12 cycles per second (cps). As I do, my own **alpha** brain waves drift me into relaxation. From our shared alpha state, I sink even more, into my slower **theta** waves of inward focus and joy (4-8 cps), then deepened to my **delta** waves (0.5-4 cps). I am on my sleep journey.

Of all of the five brain waves possible, I may spend most of my young hours predominating the dreamy, perceptual state of the delta waves.

When their eyes are starting to close, I will lower my voice even more and just make it fade away and watch...

They'll go to sleep every time."

Duermete Mi Hijito

Dee Fry
Used with permission

Alpha entrainment ♩ = 130

Duer-me-te mi Hiji-to. Cer-ra los o-ji-tos. Duer-me-te mi Hiji-to. Cer-ra los o-ji-tos.

Hear this sung at https://bit.ly/LullabyWisdom

25

The year was 1970. I was a first-year music specialist in an exquisite new school. I loved my work and the K through 6th graders. One afternoon, some excited third graders came in with concern on their faces. One was carrying a precious book with the page marked. The leader explained, "We asked Mrs. R. if we could bring this to music. She gave us permission, as long as we are careful and bring it right back. Charlotte, the spider sings a lullaby. What does it sound like?" They looked puzzled when I glanced at it and answered, "There's no music on the page, only words." They clamored in response, "There has to be music. It's a lullaby!" they insisted. "Charlotte sings it!" The class had already imagined melodies in their heads as their teacher read the poem. I mused, "Well, if we don't see the music staff, what do you think it could sound like? Perhaps we could make up the melody in our music class." And that is just what they did that day. Close to 40 years later, as I was holding my first grandchild, the melody came back to me. It sang me. – **LCS**

CHARLOTTE'S LULLABY

I love singing. Babies and singing go together like peaches and cream. My right hemisphere works at warp speed in my every waking moment.[1] 700 new synapses (neural connections) are formed every second in my first year of life. Your gentle singing has the capability to communicate soothing to me like nothing else. When babies like me in their first hours of life, had the chance to activate a soprano singing voice, they did. They cried when the singing stopped.[2]

In this study, De Casper and Cartens were looking for cause and effect responses in babies as young as 3 days old. When the infants paused between sucking, they were treated to a female singing voice. 24 hours later, researchers stopped the sound. The newborns became upset. They had "come to expect" through their organizing mind, that a singing voice was contingent on pausing from sucking. They recognized that singing was related to something that they were doing, and they could pause to enjoy a serenade. By changing their behavior, they could get the result that they wanted. How's that?

Newborns like me, even toddlers, are looking for organization, routine, and predictable **patterns** from the environment. Your faces and pleasant routines help us to feel secure. Then we put on the **'vagal brake'** which relaxes our **default danger-detecting brain** and gives calm order to our day. Thank goodness! Moreover, here's the perk. You adults like us happy, too.

Why was it so memorable? We third graders were in "**the flow**"[3] creating it within 25 minutes, easily with special, tranquil innocence. Group inspiration gave it a remarkable simplicity. As we sang, we released oxytocin and forgot to be self-conscious.[4] [See also the science explanations of #s 3, 5, 15, 26] We had a meaningful group experience[5] and gained a higher sense of the quality of life.[6] We were all enjoying ourselves. After enjoying the initial experience, we requested and serenely sang the lullaby many times over the years.

"This is the hour when frogs and thrushes praise the world from the woods and rushes."
– E.B. White[7]

Charlotte's Lullaby

Sung in 'Wind Painting' technique

Hear this sung at **https://bit.ly/LullabyWisdom**

26

[When I met Patrick Scofield and his young daughter, and told them that I was collecting lullabies, she called me the 'Lullaby Lady'.]

PS (Dad): *This is Larken.*

LL: *Larken. How old are you, Larken?*

PS: *...Are you being shy or something?*

Larken: *....Two!*

PS: *And Larken was just singing me the Lullaby Lady song.*
[And to Larken, he continued,] *That's your night song. Do you want to sing the lullaby that we sing sometimes that you have your part? Remember? That's a microphone. And when you're ready to sing your 'night nights' part, you just lean right in and sing it right into that thing, okay? Here we go...*

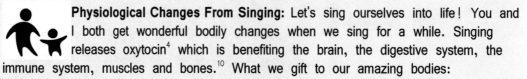

Physiological Changes From Singing: Let's sing ourselves into life! You and I both get wonderful bodily changes when we sing for a while. Singing releases oxytocin[4] which is benefiting the brain, the digestive system, the immune system, muscles and bones.[10] What we gift to our amazing bodies:

- increased Immunoglobulin A antibodies[1] (**Ig-A**) which protect the upper respiratory tract from bacteria and viruses.
- lowered cortisol (the stress hormone),[1 & 2] less anxiety.
- modulated cytokines, receptors and neuropeptides[3] favorable to healthy immune responses. We're happier with more resilient health.
- increased circulation. Increased oxygenation of the brain. Hey! As an elder, I love this one for warmer hands and feet. Keep me circulating.
- a natural "in-the-body pharmacy" anti-depressant[3] ...allowing relaxed attention.
- improved concentration and memory recollection noted for those with dementia.[5]
- more efficient muscles while singing, and stronger diaphragmatic muscles.[6]
- muscles that need less oxygen to work for the same task if we're singing.[6]
- increased anti-inflammatory, anti-oxidant and pain-relieving (analgesic) changes.[11]
- improved lung function and breathing efficiency.[13]
- increased positive social connecting, and meaning.[12]
- improved moods, released endorphins.[7] Did I hear happier? I like that.

Singing can allow us to leave behind "fight/flight/can't get it right, and enter into "rest/digest/feel your best" nerve circuits.[8]

Physiological Changes From Infant Care-giving: When I'm born, Mommy's brain changes to protect me. If a Papa is the one caring for me a lot, his brain changes too. Like a mommy, his amygdala grows 75% bigger to be vigilant for me![9] Then his meaningful connection with me can endure for a lifetime!

NIGHT NIGHT SONG

#26. Night Night Song

Music & Lyrics by: Patrick Scofield
Used with permission

Rendered in duet with Larken Vanderfield, age 2

Night Night Song

Hear this sung at **https://bit.ly/LullabyWisdom**

27

Silvi was in third grade in 1992, in Canada. That year, they were studying Remembrance Day in Socials Class. They studied the story of The Thousand Cranes and they learned how to make origami cranes. They were also part of an assembly at school and the theme was 'a thousand cranes'. It was very emotional. The kids came back from the assembly very emotional, and a little bit overwrought, and very moved by the story, the Hiroshima aftereffects, and all those kinds of things. I didn't hear this from my daughter, I heard this from her teacher who said that she herself came back to the classroom very moved and a little upset, and didn't know what to do for the kids. So Silvi went up to her and said, "Could we please sing 'Silent Night?'" So they sang 'Silent Night', and it just kind of comforted them and helped everyone settle and feel okay again. They had been making cranes the whole time. But it was using 'Silent Night' that helped them settle and feel calm again. - **IW**

****

Almost every time they go to bed, our grandchildren request to have singing. At times, they request 'Silent Night', which is a peaceful one. Even if it's not Christmas, they love it. -**VM**

****

We started over Christmas time. She was two. She sang it with the family at the Christmas Carols. She really liked the song. We added it at bedtime. It started to quiet her more than other songs. We do two verses, singing it over and over. Sometimes she would sing it with me. 'Silent Night' is one of the most lullaby-like songs we sing. It makes her feel more quiet. She's a happy child, in part, from the positive, happy bedtime songs. Even if they are quiet songs, they are happy. - **JB**

****

[JB's child narrating six years later.] *At my mom's ultrasound, my sister was 20 weeks old inside mom. She had fists that were very tight. I sang 'Silent Night' five times next to my mom's belly. It was a soft and quiet song and if I liked it, then a little baby would like it. She* [the fetus] *recognized me and knew my voice and knew everything was OK. She got calm and opened fists for the doctor. The doctor said, "I've never seen anything like this!" I got excited that my baby sister recognized my voice and she got calm.* - **BB**

SILENT NIGHT

I really appreciate it when you take my suggestions. I feel 'seen' by you. I feel 'felt', acknowledged and valued. I am contributing to my family and my class. This helps me to develop **self-esteem**, and happiness. Life's energies flow through me in inspired ways. These insights are my child gifts to you.

Safety allows me to develop my whole brain more completely with a basic supply of nutrient and oxygen-rich blood, which is building the branching neuro-synapses of my right and left hemispheres, along with the "kindness" wiring to my **middle PreFrontal Cortex (mPFC).** My PFC circuits grow, developing in me, an internal emotional balance, my empathy, compassion, response flexibility, and fear reduction. I gain insight, my moral perspective of a larger good, and self-knowing awareness, as well as my ability to plan and problem-solve. All these areas of my brain need my perceived safety and in-person responsive face-time to wire up together, and integrate their signals.[1]

Silent Night

Franz Gruber/J. Mohr
Sung by Jennifer Bolton

Si - lent night. Ho - ly night. All is calm. All is bright

'Round yon Vir - gin Mo - ther and Child. Ho - ly In - fant so ten - der and mild.

Sleep in hea - ven - ly peace.___ Sleep in hea - ven - ly peace.___

Hear this sung at **https://bit.ly/LullabyWisdom**

28

SLEEP WELL (GRANDBABY'S LULLABY)

What a beautiful cold winter's day. Sun and clouds, Cascade peaks and Olympic mountains brightened up and framed the dark gray waters of Puget Sound. Along with the morning crow and chickadee calls, we have just enjoyed 50 minutes of piano lap time, singing lullabies and nursery songs with old family songbooks. I am fixing breakfast for her older sibling. As she snacks in her high chair, 21 month-old Bebel begins singing over and over, "Sleep well, sleep well, sleep, O my baby." At first, only 'happy baby' registers in my mind. After her fifth repetition, I realize she is creating. I sing it back to her. We trade phrases back and forth. Then she changes two pitches of the line. I echo the new phrase back. She repeats it, and again I echo her melody. Bebel adds the phrase, "My mama's gotchoo." I laugh. It's a happy memory cozied up in the sleepy song. Then she's done and counting oaty O's with a pudgy finger.

*She is fond of playing **"I'm gonna get you!"** which ends after my tiny pretend-chasing steps, and a gentle grab with the word, "Gotcha" or sometimes "Gotchoo!" to the toddler's squealing delight. What a happy nap sentiment: "My mama's gotchoo."*

After lunch, I shared the song with her sister, and Bebel added, "baby lullaby," then toddled off. "Sleep Well" is her song of choice for naps now. -LS

 My sister sang a lullaby (#27) to me when I was a fetus during my 20-week in-utero ultrasound. From my prenatal life, I have been internalizing soothing sound. I am beginning to form the chemistry and behaviors of the **CARE brain circuit.**[1] I am imitating behaviors in which my family expresses motherly and fatherly CARE. While I sing my newly-forming 'baby lullaby', my self-soothing voice is producing some of the the CARE circuit chemistry, oxytocin. Through the epigenetics of a CARE-filled family, I am epigenetically gaining gene expressions for increased **resilience** in the face of stressors.[2] I am gaining:

- more receptors in my brain for **glutamate**[3] (ready to learn)
- more **norepinephrine**[4] (mobilizing body and brain, also for my focused attention)
- more receptors to take in **GABA**[5] (relaxation & sleep onset)

PLAY[6] **brain circuit:** My punch line in the song is a memory about rambunctious yet gentle **playtouch**[7] and activating my PLAY limbic (mammalian) brain circuit with those around me. Chasing[8] is just one form that activates the PLAY circuit and gains me social joy.[9] PLAY only occurs when I am safe, secure and feeling well.[10] Physical PLAY is part of my mental well-being[11] and also builds my social confidence. I give a lot of eye contact to you during the fun. PLAY-ing, I also produce:

- a widespread release of brain-produced **opioids**[12] (euphoria)
- some **dopamine**[13] (anticipation)
- endogenous **cannabinoids**[14] (playfulness, giggles)
- the gene expression **NR2B**[15] (learning & memory formation) in my frontal cortex and amygdala.
- **Brain-derived Neurotropic Factor** (BDNF) which promotes neuron-sprouting in my hippocampus[16] (memory).
- **Insulin-like Growth Factor**[17](IGF-1) growth hormone for maturation ('brain fertilizer')

My first 2 years are a time to absorb my biological set point for happiness.[18]

What a happy nap sentiment: "My mama's gotchoo"

Sleep Well

Bebel, 21 months
Sung by Amah, Poppy, & Tom Caruso

Sleep well, sleep well, sleep o my ba - by.

Sleep well, sleep well, sleep o my ba - by.

Sleep 'L sleep 'L sleep o my ba - by.

Sleep "L sleep 'L My ma - ma's 'got - choo'

Hear this sung at https://bit.ly/LullabyWisdom

29

This lullaby came out of the necessity of the moment. Little Bebel wanted to be asleep and wasn't. She started to fuss. I began singing and holding her with a gentle motion. Her ten-year-old sister joined me, looking into Bebel's eyes as we sang. Baby began to yawn and lower her eyes. Her older sister spontaneously made up more verses. We both did until Bebel was very sleepy, then we quieted to a hum. Over and over... **-LCS**

Singing and eye-gaze are simple ways of passing on soothing skills to another generation. When you include a sibling like me, you honor my gifts of love and care for my younger sister. This raises my self-esteem, helping me to see myself as capable, on my way to adulthood. Humans are "wired for WE".[1] We light up with signals in most of our brains when we are interacting with each other. Together we provided a positive model for relating and creating. Grandma and I connected through Baby's fuss; together, we lulled her to sleep.

FACES RULE! I have three pathways of Vagus nerve to keep me safe. **My three Vagal nerve circuits are in hierarchical relationship.**[2] I prefer looking at you with my Ventral Vagal Social Engagement System circuit (SES). That is just wondrous. I can only use it when I feel safe. Our caring eye contact tells me I'm safe. We are producing healthy, protective oxytocin. Then I am also using my Parasympathetic capabilities of relaxing, digesting, and optimally building my brain with you to be a complete, sociable human. Our social biology protects us, and positively changes our brains continuously, to connect with ourselves, each other, and our world.

However, I am born with an continuously active, alert system turned on. It is instinctual with me. It's called **neuroception**[3]. I unconsciously hunt for cues of safety or danger. If I feel danger from outside me, (too cold, too loud) inside me, (hungry, tired, hurting) or from another person, (too unfamiliar, yelling, crabby, not available to interact with me, distracted) I may instinctively and unconsciously activate my Sympathetic Nervous System (SNS) and my brain's amygdala signals the alarm. I've got to get out of discomfort. If SNS fight/flight fussing and crying are not enough to remove my danger, then in extreme circumstances, I instinctively activate the old Dorsal Vagal for immobilization/death-feigning, as a last resort to protect my internal organs.[24] (That's my emergency preparedness package.)

You big humans love to see warm faces, and connect to others. Many of you are even connecting globally. Yet for toddlers and babies like me, it is *you:* the face with the cuddles and closeness that soothes and grows me the healthiest for the first three years. Together you and I are growing my health, smarts *and* kindness.

Motion strengthens my immune function by stimulating my infant, passive, lymph circulation. With babies like me, it is often a very gentle, tiny bouncing that many adults instinctively and repetitively provide.

*"for babies and toddlers, it is **you**: the face with the cuddles and closeness that soothes and grows them"*

Hold You Close

Sing any or many or a few verses in repetition,
or none....just humming into the expansiveness of intimacy...

Bijou B. & L.C. Seaman
To Bebel 4/10/2018

2. Circling 'round the starry skies (3x)
   Gifts of twinkling humor for your eyes.

3. Floating on the Sea of Dreams (3x)
   Ebbing, flowing, up and down the streams.

4. Close your eyes and drift away. (3x)
   When you awake we'll have some time to play.

5. For tomorrow's another day (3x)
   We can share some more with lots to say.

6. Sleep, sleep, sleep, don't you cry (3x)
   While we sing to you this lullabye.

7. Twinkle Toes, it's time to rest. (3x)
   When you wake, patty-cake and feel your best.

8. Lie on down and rock to sleep. (3x)
   Counting fields of fuzzy fuzzy sheep.

9. Bebsie* Boo-oo-oo-oo-oo (3x)
   Listen to the owlet call "Who who".
   * *Your child's name here*

Hear this sung at **https://bit.ly/LullabyWisdom**

30

◀◀◀◀ *LULLABY SUPERHERO* ▶▶▶▶

That was always part of the magic. Stroking the head lightly, just forward and back Slow and gentle. ... with a little flutter to it, so the speed going down the head would have been slow, but it had a little bit of that flutter. ....just a light tapping....

I think it's the touch. I think I could do that with a person without singing, and calm them. I used to do it with my mom. I calmed her when she was demented. I talked with her constantly. Not a massage, but just stroked her arms [John's whole hand moves up and down the arm, tapping with spread fingers as a group] *...just talk to her in that calm, slow way and the "tendering. I do know that I have a very gentle voice. She would go from being rigid to relaxing.*

Nate always needed to be walked. So I walked and cuddled him. It was all part of a package. There was my movement, my walking. There was my holding, there was my stroking, and then the song. Mostly I remember slowing down as I could feel him relax and be on the verge of sleep.

And since then, if I see a child screaming in an airport or a restaurant, something like that, I don't do it often enough, I'm too embarrassed, too proud, too vain, but if I can get over that, I'll just say, "Do you mind if I just held your child for a bit while you finish your dinner?" or "Would you like a relief?" And I know within fifteen minutes I've got them calmed. Just by walking and talking and patting and holding their head. It's so important. Sometimes just holding the head there. Seeing them going to sleep.

L: You've got man-sized hands holding a little head. Looks like a softball catch.

Yeah, it is a hemisphere for a baby's head. The last time I did this was at a...oh, it was a beautiful place. It was Hurghada, a resort town in Egypt. There was a couple; they were European. We were at a lovely restaurant out on top of the... exquisite food and everything else. This young couple with a baby, maybe eight months, less than a year. It was fussy, and fidgety and doing stuff. The mother was constantly having to pay attention to the child and they were not paying attention to their meal. So I thought, what the heck, and I went over to them. "Would you like me to take your child for a little bit so you can eat your meal?" And she looked at me, and she had no doubt or anything, and she said, "Oh, how wonderful." So at first I walked the child in the stroller. Then I picked him up, put him over my left shoulder and started doing the same thing, just walking and talking. With the hand patting in fluttering. And I did it for about a half hour and they had finished that meal. Then I went back to Mary Lyn and we finished our meal. It was just great.

Grandkids? I've used the same song with [grandson] *Kedar. Basically it was the same. I figure it works, why change it?! Asleep in five minutes, usually.* -**JV**

Stress Relief: In the arms of a soother, one relaxes. The dining parents' anxious brains switch to their happier, PNS nerve pathways, allowing them to relax and enjoy their meal time with optimum digestion.

Fluttering: According to Chinese xi-gong medicine, when the hand and arm are fluttering, the motion brings a lightness and open joy to the heart meridian.

ONE AND ONE

"It was all part of a package. There was my movement, my walking. There was my holding, there was my stroking, and then the song."

One & One

John Villaume
Used with permission

1 and 1 are 2.   2 and 2 are 4.   4 and 4 are   8 and I don't   know an-y more. So

sleep babe,   sleep my   babe. So   sleep babe,   sleep my   ba - by.

Hear this sung at **https://bit.ly/LullabyWisdom**

31

I just had found out that my two month-old grandson was in the hospital with two kidney infections. He did not have a certain anatomical structure developed at birth to keep his kidneys protected. I started to worry. He had been born six weeks premature. He had great care. Yet I still had fears and worries (mid-brain circuits.) When I was with him, I needed to be whole-brained to offer him strength, perhaps playful or entertaining distraction from his physical discomfort. The song came to me a capella, then to my harp. I sang it and sing it to remind me that he will live as fully as possible with the whole-brained support of all who wish him well.

To help keep worry and fears at bay, I unconsciously enlisted the long exhalation of a song phrase. The phrase of over 8 seconds flipped my nervous system out of the FEAR circuit of the limbic 'downstairs brain' and into the neocortex 'upstairs brain', accessed by the Parasympathetic Nervous System.[1] I grew calmer. Then I could access whole-brain inspiration to reinforce my love and compassion for his best outcome. Because this young one learns hypnogogically, he absorbed my process when he was with me.[2]

♩♫♩♫

"When he was 21 months, he wanted a nap. He'd ask for 'baba' (a sippy cup of warm milk) and cuddle time, leading me to the rocker. After milk, he'd tug me to the front door: he wanted to be held outside. I carried him. We said, "Night night, Maple", he patted the tree trunk, and other trees in turn. We came to rest under the welcoming cedar. I shifted from one foot to the other, back and forth, and back and forth. He slowed to a relaxed body. His eyelids softened and began to close. It our last tree to remind that it was nap time."

It turns out Japanese and Korean biochemical researchers have been studying these effects for decades, naming the phenomena 'Shinrin-yoku' or **'forest bathing'**.[3] Within 16 minutes of being among trees, our human cortisol stress levels drop, our blood pressure lowers to a healthy rate,[4] and our pulse rates stabilize for nourishing parasympatheic nerve activity.[5] Exposure also improves our sleep quality,[6] strengthens our immune systems,[7] lung capacity,[8] and the elasticity of our arteries.[9] **Baby benefits from being a bit 'round the bark and greens. Even 20 minutes of forest time increases our ability to focus, to improve our attention and to control cognition.[10] Take that, ADD!**

What is the cause of the baby's physiological changes? Trees act as huge air filters removing tons of toxins.[11] Tree bark contains chemicals, called **phytoncides**, to protect the tree from disease. Phytoncide molecules are airborne, as well. Biochemists found the isolated phytoncides benefit us as well as the trees.[12] Baby breathes in phytoncides and relaxes. To make it more fun, rub fir needles or cedar fronds between your fingers and place near baby's nose and/or your own. Sniff the calm. Even breathing the humus scent from the forest floor causes **oxytocin** release in humans: from our own Mother Earth.[13]

JULIAN'S LULLABY

Julian's Lullaby

Licia Claire Seaman
2009

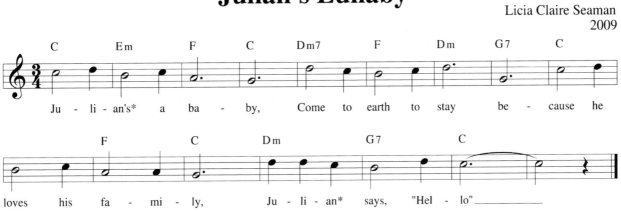

Ju - li - an's* a ba - by, Come to earth to stay be - cause he

loves his fa - mi - ly, Ju - li - an* says, "Hel - lo"_____

*Your child's name here.

Hear this sung at https://bit.ly/LullabyWisdom

32

COMPASSION LULLABY

May Our Hearts Enfold You

Sometimes our lives take unexpected turns, and sometimes grief saps us. Our blessing is that we have each other, in families, communities, and those who wish us well. When tears flow, we can sink deeply into pain pools. Yet, sometimes our mindfulness gives us another emotional color alongside the grief. 'Compassion Lullaby' allows our voices to soothe and comfort us through the sorrow walk.

I have used it after the loss of an infant, sung it when saddened by the progression of a friend's dementia, and later hummed it to calm a confused memory care resident. I use it as a meditation, a focus, a vehicle to send love to relations after a sudden death, and to remind me that our humanity always has the option to include kindness in our response to loss. Even as life events upend us into helplessness, we can gently step into being the gift. Then comfort each other and be available.

When we sorrow, we enter into the GRIEF (loneliness) circuits of the mid-brain, the limbic (emotion circuits) area. If our Pre-Frontal Cortex (PFC) areas are well-developed by care and mindful practices, we also may have the option to touch the sorrow, then ask. "What else is a possible response that can soothe us and others during the pain? Through the ache, what can we bring?" In this case, *May Our Hearts Enfold You* answered me. My creativity reengages the neo-cortex through the CARE circuit in the limbic brain. The humming, whistling or singing of long phrases puts the 'Vagal brake' on my fight/flight Sympathetic Nervous System pain pathways, and switches me to the rest and restoration of Para-sympathetic Nervous System pathway. Yes, in the early days of grief, I found myself humming or whisper-singing the song, just to continue my necessary obligations. I still return to the lullaby when I chance to fall into sorrow again.

If you are grieving and have a young one like me, remember that emotions are contagious.[1] Lullabying and cuddling may soothe me during sadness, releasing oxytocin which stimulates the release of endorphins and dopamine.

*"Nurturing mothers experience **major endorphin surges** as they interact lovingly with their babies − endorphin highs can be one of the natural rewards of motherhood."*

–Gabor Mate

Daddies who are lead care-givers also experience this. Dopamine is a brain 'feel-good' chemical. It's part of anticipation. You get elated with "a sense of infinite potential." It is important for motivation, incentive, and energy.[3] Oxytocin is also a soothing part of the limbic system[4] and affects your moods as well.[5]

[The brain's internally-formed, endogenous] *"opiates are the chemical linchpins of the emotional apparatus in the brain that is responsible for protecting and nurturing infant life."* [6]

–Gabor Mate

Our rewards are oxytocin, endorphins, and dopamine neuro-transmitters. When you are around me, you and I feel them. I'll help.

Compassion Lullaby
(Adapted from a Buddhist prayer)

Licia Claire Seaman
May 5, 2017

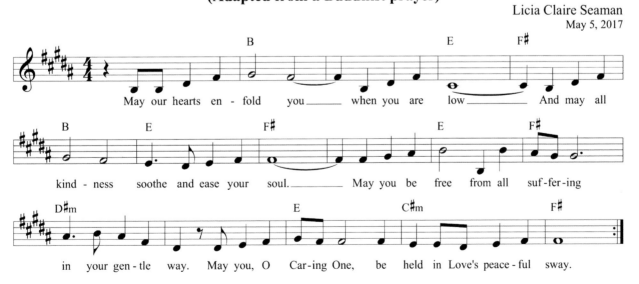

May our hearts en-fold you____ when you are low____ And may all kind-ness soothe and ease your soul.____ May you be free from all suf-fer-ing in your gen-tle way. May you, O Car-ing One, be held in Love's peace-ful sway.

The words of *Compassion Lullaby* are optional after the piece is familiar to you.
Humming carries the essence as well, and can create a simple lullaby or soothing moment.

Hear this sung at **https://bit.ly/LullabyWisdom**

Part II: Simple Soothers

33 Walking Rocking: *"step – rock"*

Inside or in the backyard....in a little quiet place. We go step by step slow-ly and we're just rocking the baby down and up...

You take a step and then you rock back. You take another step and then you rock back. And take another step, and then Left...then Right...[while stepping] you rock forward. Then back... step and rock forward. Then back....just slower and slower. I am moving with it and slow-ly and slow-ly...

Then let the baby down from your back and lay her down slow-ly and let her sleep. And if she wakes up a little bit, then start singing again. One more time, until she goes back to sleep.

Acholi Song Translation (from Sudan):

> *Mother is singing and is going to put the baby to sleep.*
> *She needs to cook.*
> *She was cooking and then the baby started crying.*
> *Now she has to wrap the baby so the baby can go to sleep.*
> *Then she will come back and cook for the family*
> *And that's when everybody will be happy...*
> *And the baby will be happy, too.*
> *When the baby is sleeping nicely, that's when the baby's happy.*
> *And then the mama is happy, too*

You have to do so much work. I always play with them to get them tired before putting them to bed. I always read to them. I sing other different songs. They like that. I sing one song. Then another one.
— ***Estella Amwony***
Mother of three capable children

34 *'Cushla, cushla, cushla cushla'*

*Ma Cushla means, 'My Love' in Irish. [The Gaelic phrase '**mo chuisle** ' literally is 'my vein, my blood'.] I was surprised one day at S G's house that she had a bear, a little bear that made the sound of the heart and the circulation 'in-vivo'. That would put Hope to sleep ...that same '**Cushla, cushla, cushla**' sound. I tried it when I was babysitting for Hope. She was getting a lot of teeth at once, and she wasn't sleeping well. She was really tired, but she wanted to play, and so she was just fighting going into sleep. I had sung her the little pie-baking song. It's entertaining and she wants to stay awake for that, but after that, I just rocked her and said, 'cushla, cushla, cushla' and her eyes rolled into sleep and she was gone...*

...It's nice to fade out a little bit. It feels so loving...calling her 'My Love' while I'm rocking her. Nice and secure. I don't think I got more than three times, then she was gone. I kept at it for a little while to get her into a very deep sleep, and held her while she was sleeping on my chest. She'll be a year and a half this month...having a bad day.
— **Kelli Calderwood**

SIMPLE SOOTHERS

Hear these soothers at https://bit.ly/LullabyWisdom

35   For preemies: *'sip sip sip'* [inhale]

Repeat it gently – pace it at a comfortable pulse – inhale the same length as a 'sip'.

This Simple Soother is made for the infant with very sensitive perception. Maybe, it is for a baby who is extra fussy as well. If you feel drawn to it, try it. It's all about the soothing. This technique is from Connie Kirtan, second-generation hospital maternity nurse. Her mother also was a maternity nurse!

36   *'sh hh   sh hh   sh hh hh'*   [2+2+3 seconds]

This sound, like that of waves and waterfalls, is developmentally supportive to babies and anxious ones. Repeat until sleep is secure. Each pulse is 67 BPM.

It's slow. It's soft. It's sweet. It has repetition. It didn't require interest. I didn't do it with any of my others. Somehow it came to me with Elora at nine months, my fifth grandchild.   **– Marian Grebanier**

37   Whispering in Baby's Ear I: *'shi shee___   shee shoo___'* [repeat until asleep]

That's how I put my babies to sleep. I wrap the baby and hold them up to my cheek. I start singing to them. I hold them up to where my jaw is so they are close to my voice. **shi shee____   shee shoo____** [100 BPM] **– Rita Long-Visitor Holy Dance**

Rita Long-Visitor Holy Dance is an Oglala Lakota Great-great-grandmother.

Used with Permission. Recorded excerpt from **Sacred Blessings CD – Native Songs Stories & Lullabies. A collection of indigenous elder wisdom for the next seven generations and beyond. International Council of Thirteen Indigenous Grandmothers.** 2009 Samadhi Life, Inc. All Rights Reserved. www.samadhilife.com

38   Rocking Circles With Upper Body

The baby is behind you, on your back in a wrap-around cloth. You are rocking, just calming... not too fast. You go....making circles, going very slo-ow, so the baby doesn't awake. (One rotation may take 20 seconds.)

While standing, Veronica starts the rocking, slightly bending from her waist, so slowly, then around. Her upper torso makes the circles, leaning forward, rotating to the side, then to the back and all the way around to the front again...like drawing a horizontal letter "O" with her chest. She leans into the front, then the side, the back, and to the other side, and over and over.   **– Veronica Ded Dei Tetteh**
From Ghana tradition
Used with permission

39 Descending Hum: *'Mmmmmₘₘ'* [for 2-3 seconds]

I wanted to replicate a sound that was emulating calm: going from anxiety to calm. I most often used this when she was crying or upset in the middle of the night, to get her to come down to being relaxed and feeling soothed. So I didn't want a particular word, just a sound I could repeat over and over again while I was rocking her in the rocking chair.

Then I'd take my hand, put it at the top of her forehead, bring it down, then over her nose and bring it off her nose. When I did it, it would be covering over her eyes, so that way, she'd want to close her eyes and go back to sleep in the middle of the night.
— **Jennifer Bolton**

JB started with the pointer finger down the nose, while keeping her fingers together. As she progressed, it was the middle finger over the nose, still keeping fingers together, moving the hand as one. A physical therapist may encourage one to sing similar, repetitive, descending, sliding glissandos. Such sounds relax the muscles.

40 Wind Painting I

The melody is also on Unci Rita's *Sacred Blessings CD* as "Whispering in Baby's Ear II" and then from Spirit through Licia as well.

Wind Painting refers to a vocal style of singing: using pitch and creating melodies with the 'shh' sound, by changing the vowel in the mouth and the lips. This is one example. Feel free to create and follow your own inspiration. The soft sound is very intimate. The present is soft. (The slow phrase can last 7 ½ - 8 seconds.)

Shhh_____   Shhh_____

41 *'shi-shi-shi-shi shhhhh'*

Now with my grandchildren that's a little more recent memory. The oldest is going to be a high school junior and the youngest is 2 months. And of course, the memory of the two-month one is probably the strongest. I can say this about any of the other grandchildren, that, if they were fussing, I simply said to the parent, "Let me give it a try." I would put them on my left shoulder and place the right hand low, left hand high and we would just walk around in the house, slowly. The head would probably be on my shoulder, if it wasn't still bobbing around. It seemed to settle them down, just the walking, and then I would talk,

SIMPLE SOOTHERS

It's quiet time. ... I would just whisper it to them.
That's all right. That's all right. You're okay. You're OK
shi-shi-shi-shi shhhhh
You're OK....you're all right....yeah.
shi-shi-shi-shi shhhhh

— **Doug Anderson**

42 Wind Painting II [~ 6 - 6 ½ seconds a phrase]

Shhh_____   Shhh_____   Shhh_____   Shhh_____

— **Licia Seaman**

43 Motherese

*I went over and talked to her in the gently rising and falling voice of 'Motherese'.
After 10-15 seconds of my voice, Jensyn stopped crying and just looked at me,
studying my face. It's so much fun! I kept it up! My son, her dad, came in and said,
'Hey, how did you do that? Why doesn't she do that for us?'* - **Molly Zimmerman**

44 Calming the Fussy Hour

I'd find any little postage-stamp-sized park I could, anything. And she would relax.

*I would take her out in the dark. I remember one specific day. This is when my
daughter had a house. We were out under the foliage and the trees and with the
squirrels. I took her out at night, 4:30 or 5:00 p.m. when it was the 'fussy time': the
hour before mom returned home. It was winter, so it was getting crispy. I'd get her
in the snuggly carrier near my heart, all bundled up, then head out and do* **'smell-
o-rama'.** *We'd go up to a pine tree. I'd rub the needles until my fingers got all piney,
and put it near her nose. She kind of enjoyed that. Then I'd try rosemary. Next, I'd
try a little Melissa Officionalis (lemon balm), that tingly, tangerine kind of smell
and put that lemony smell, that lemon balm, near her nose. Around five blocks
away, I got to one tree, an evergreen, and I can't remember who it was now, but I'd
love to go see it at that old address again. When I got near, it had a particularly
soothing smell: pine but not sharp. Calming is what I felt when I smelled the
boughs. When I put it under her nose, there was this big sigh that released from
inside her, tiny as she was, and I thought, 'Memorize that plant.' And I did.
Whenever we went out to walk, I made sure I got that one."* - **LCS**

Hear these soothers at **https://bit.ly/LullabyWisdom**

This Grandmother knew that my brain's **amygdala** (fight/flight alarm), is responsive to trees. The tiny almond-shaped brain structure designed to signal perceived danger, lies within my limbic circuits for emotional responses. My amygdala calms when it sees **tree foliage**.[1] It can signal my entire nervous system from my prenatal beginnings to my 'wellderly' (elderly) years. I am also smelling and enjoying the fragrances and phytoncides cited in #29 and #31.

Later, when I was a chewing little person, we would eat one or two sweet rosemary blossoms after asking the plant's recommendations and permission for harvesting. Yum. Rosemary increases mental clarity and relaxes my body, reducing stress hormones by 23%.[2]

I also like to join in **'Smell-o-rama'**. If we are outside, you can rub a fir needle, cedar leaf, juniper needle, lavender stem, or rosemary needle between your fingers, then place your fingers an inch below my nose. Or show me how to sniff flowers, or rose petals. Ummm. The herb and conifer smells lower my cortisol levels.[3] They relax my body[4], and ready me for sleep.[5] Ummm...

45 Grandfather's Walk

We lived in a small town in the south part of Chile, so it was forests and rivers, ...more like the countryside. Back in the time, it was so dark always and everywhere, we used to go with my grandfather and take a walk before we went to bed during beautiful summer evenings where the sun set and then become dark. We used to go see all the fireflies.

We went walking with him. Then on the way back, he would hold me and put me on his shoulders, because at that moment I was just ready to rest. My body was tired and he used to carry me on his shoulders and bring me back home.

Sitting on his shoulders, of course, I would just play with his hair, his white, silver hair. It was very unique.

The walk in the evening before bed is another thing you can do with the kids. And it's very relaxing because you know, all the energy goes in the walk and you relax, and you come home. And so I have all those memories. They were very precious for me. It transports me to another world. I just go back in the time and....so relaxing ....such a good memory.

— **Yolanda Rubio**

46   Trees 'N Sleep

By mid-spring, the trees across the driveway were leafed-out in cooling light-green foliage. They were visible from the bedroom. At nap time, four year-old Julian R. said, *"I love looking out the window at the trees. It's so beautiful. Even when I sleep, I'll be looking out the window."*

47   Nature

Grandmother says, *"He was fussy. His little limbs were getting restless. His baby talk was starting to sound anxious and over-active. His eyes were gravitating to this and that, with no time to enjoy the sights. His ears were getting more sensitive to any noise, not just my vocal exploration. and our conversation. Sleep was next on our routine, but 'Little Squirty' was too over-stimulated to enjoy the slow lull of the lyrical rocker. I carried him in my arms outside. We went around the block. If his eyes stopped to focus on a branch or bird, I paused. Perhaps my patient attention let him attune to the slower natural surroundings. Then, wind in the trees overflowed in soothing whispers. The raindrops on piddly puddles tantalized his waning eyes. The glowing flowers and scents of trees drifted into his inhalation. Sometimes I assisted by taking a pine needle, smooshing it between my fingers and positioning it just below his nostrils for a few breaths. Before going inside, we stood under a cedar. He glanced up at the spider's silken mandala, then to the wind in the boughs. In a few moments, his limbs calmed and quietness massaged his mind. Nurturing nature lulled us both with her comfort."*

48   ...and, remember to drink water.

Our neurons fire faster.
Our lungs are more efficient.
Our attention is clearer.
Our memories are present.
We are less anxious, calmer with water.[1]
The voice is muscle...good singers pee freely.

ENDNOTES

Introduction

1 Mate, Gabor. (2010) *In the Realm of Hungry Ghosts.* Berkeley: North Atlantic Books. 175

2 Ibid.176.

3 Ibid, 191.

4 Dawson, K.W. & Fischer, G. (1994) *Human Behavior and the Developing Brain.* New York: Guildford Press, p.9.

5 Mate, Ibid.

6 Siegel, Daniel J. (2012) *Pocket Guide to Interpersonal Neurobiology.* New York: W.W. Norton & Co. p. 27-2

7 WNYC's Radio Lab with Jad Abumrad and Robert Krulwich delving into the making of their acclaimed program that melds science, philosophy, and psychology this podcast "short" from November, (2007),episode "Sound as Touch" with Dr. Anne Fernald, Director, Center For Infant Studies at Stanford.

8 Janata, Petr. And Birk,J. et al (2002) "The cortical topography of tonal structures underlying Western music." *Science* 2002 Dec 13:298 (5601): 2167-70. DOI: 10.1126/science.1076262

9 Janata, Petr. (2009) "The neural architecture of music-evoked autobiographical memories." *Cerebral Cortex* (epub. Feb 24, 2009) 2009 Nov 19 (11) 2579-94.

10 Diamond, Marion Cleeves. (1988) *Enriching Heredity.* New York: The Free Press/Simon and Schuster.

"The reputed pediatrician, T. Berry Brazelton, points out that infants exposed to too much stimulation respond either by crying, by extending their periods of sleep, or by developing colic or withdrawing from any new approaches. In providing increased stimulation for the young, the adult, or the old, one always has to keep in mind the need for adequate time at each phase of information processing: input, assimilation, and output. The integration of the input is essential before we can anticipate a meaningful output. As adults, we frequently say, "Let me think things over." It is essential to give the infant the same opportunity."
http://www.sfn.org/~/media/SfN/Documents/TheHistoryofNeuroscience/Volume%206/c3.ashx

11 Potkin, Katya Trudeau. and Bunney, William E. (2012) "Sleep Improves Memory: The Effect of Sleep on Long Term Memory". *PLOS.* https://doi.org/10.1371/journal.pone.0042191

12 "Sing Your Way to Fitness" Sci Mind May/June 2014 , p.17, and

13 Fritz, Thomas et al. (2013) "Musical agency reduces perceived exertion during strenuous physical performance", PNAS. 1217252110 [...and besides, it is easier to stay in pace with the beat.]

14 Carter, C. Sue. (2021) The Healing Power of Love: An Oxytocin Hypothesis. Kinsey Institue. Indiana Univ. Delivered at *Solving Chronic Pain Summit* Jan.22-24. Open Center, NY

15 Crowley, Chris & Lodge, Henry s. (2004) *Younger Next Year.* New York: Workman Pub.

16 Porges, Stephen. (2017) *The Pocket Guide to the Polyvagal Theory: The Transformative Power of Feeling.* New York: Norton

17 Crowley Ibid.

18 Ibid

19 Ibid p.255

20 Porges, Stephen. & Dana D. (2018) *Clinical Applications of the Polyvagal Theory: The Emergence of Polyvagal- Informed Therapies.* New York: W.W. Norton.

21 Kolb, Bryan. Mychasiuk, Richelle et al. (2012) Experience and the developing prefrontal cortex. PNAS Oct.16,   2012 vol.109 supl. 2 p.17186 – 17193.

22 Ibid

23 Carter Ibid

24 Kolb Ibid

25   Siegel, Daniel J. (2007) *The Mindful Brain*. New York: W.W. Norton.

26 Kolb Ibid

27 Porges & Dana (2018) Ibid

28 Ibid

29 Tanaka-Arakawa Megumi M., Matsui, Mea. et al. (2015) Developmental changes in the corpus callosum from   infancy to Early adulthood: A structural magnetic resonance imaging study. Plos One 2015. 10(3): e0118760.

30 Porges & Dana. Ibid

Part I: Cuddles and Cradle Songs

1. Sunrise Song

1 Perry, B. & Pollard, R. (1998) "Homeostasis, Stress.Trauma, and Adaptation: A neurodevelopmental view of childhood trauma." *Child and Adolescent Clinics of North America 7* (1) (January 1998):33-51 Citing data from R. Shore, (1997) *Rethinking the Brain: New Insights into Early Development*. New York: Families and Work Institute.

2 Ainsworth, M, Blehar. M,Waters. E.,et al.(1978) "Patterns of attachment: A psychological study of the strange situation." Hillsdale, NJ Erlbaum.

3 Thank you Margaret Rose Bolton and baby Viola Lum.

4 Yuan, Lin et al (2016) "Oxytocin inhibits...inflammation in microglial cells..." *Journal of Neuroinflammation*

2. Sweetest Little Child

1 & 3 Carl Sherman . (2011) "The Neuroscience of Improvisation"

[See more at: http://www.dana.org/News/Details.aspx?id=43160#sthash.sFJoq0N3.dpuf]

Cognitive neuroscientist Jamshed Bharucha, of Tufts University, spoke similarly about creative spontaneity. "You can't improvise if you haven't been engaged in learning, practice, guidance, discipline—a lot of hard work. There's a myth about creativity as magic. As a scientist, I don't believe in magic, but in years and years of mastering what's already there, so you can go beyond it." - See more at: http://www.dana.org/News/Details.aspx?id=43160#sthash.sFJoq0N3.dpuf

Aaron Berkowitz of Harvard University said, "The only difference we saw was that for some kinds of improvisation, musicians were turning off an area of the brain, and non-musicians were not: the right temporoparietal junction, which is…typically deactivated in situations of goal-directed behavior" to inhibit distraction by irrelevant stimuli that might impair performance. He speculated: "I could imagine that when he came in and started to play, he shut everything down and was one with the instrument."

2 Limb CJ, Braun AR (2008) "Neural Substrates of Spontaneous Musical Performance: An fMRI Study of Jazz Improvisation." *PLoS ONE* 3(2): e1679. doi:10.1371/journal.pone.0001679

Charles Limb, a Dana Foundation grantee who teaches at both Johns Hopkins Medical School and the Peabody Conservatory of Music, has used fMRI to look into the brain of the improviser. "Music affects the brain globally," he said at the symposium. "We were trying to figure out what part of the brain is involved specifically when you improvise."

3 Limb, Charles. (2008): "Figure 3. Three-dimensional surface projection of activations and deactivations associated with improvisation during the Jazz paradigm." doi:10.1371/*journal.pone*.0001679.g003. Medial prefrontal cortex activation, dorsolateral prefrontal cortex deactivation, and sensorimotor activation can be seen.

4 Mindset Works Inc (2002-2013) p.17.

3. Five Generations of Lullabies

1 C Sue Carter and Stephen W Porges (2013) "The biochemistry of love: an oxytocin hypothesis." EMBO Rep. 2013 Jan; 14(1): 12–16. Published online 2012 Nov 27. doi:10.1038/embor.2012.191] https://www.ncbi.nlm.nih.gov/pmc/articles/PMC3537144/#b7

2 C Sue Carter (2021) The Healing Power of Love: An Oxytocin Hypothesis. Kinsey Institute, Indiana University

4. Tum Tum Tum - Apague A Luz (Brazilian Lullaby)

1 Crittenden, Patricia McKinsey; Landini, Andrea (2011-06-13). *Assessing Adult Attachment: A Dynamic-Maturational Approach to Discourse Analysis* (A Norton Professional Book) (p. 186). Norton. Kindle Edition. *Analysis by Mark Baumann.*

5. Grandmother's Lullaby

1 Xie et al "Sleep initiated fluid flux drives metabolite clearance from the adult brain." *Science*, October 18, 2013. DOI:: 10.1126/science.1241224

2 NIH news releases Thursday, October 17, 2013 2 pm EDT "Brain may flush out toxins during sleep. NIH-funded study suggests sleep clears brain of damaging molecules associated with neurodegeneration."

3 Iacoboni, M.. et al. (2004) "Watching social interactions produces dorsomedial prefrontal and medial parietal BOLD fMRI signal increases compared to a resting baseline." *Neuroimage* 21, 1167–1173

4 Uddin, Lucina Q.,Iacoboni, Marco, Lange,Claudia and Julian Paul Keenan. (2007) "The self and social cognition: the role of cortical midline structures and mirror neurons." *Trends in Cognitive Science.* Vol.11 No.4

5 Carter, C. Sue. (2021) The Healing Power of Love: An Oxytocin Hypothesis. Kinsey Institute, Indiana University. Delivered at *Solving Chronic Pain Summit* Jan.22-24. Open   Center, NY

6. Your Mama Has To Be Gone Now

1 http://www.ted.com/talks/alison_gopnik_what_do_babies_think/transcript?language=en

7. Tu Tu Teshkote – Aztec (Nahuatl) Lullaby

1 The song "Tu Tu Teshkote" was from the CD *Under the Green Corn Moon* (1998). It is titled "Aztec" and sung by Lorain Fox of Cree/Blackfeet ancestry. She learned this beautiful Nahuatl lullaby from her teacher, Maestro Tlakaelel. Kathryn Langstaff learned it by ear from the CD.

2 http://www.nature.com/srep/2013/131023/srep03034/full/srep03034.html Hidenobu Sumioka, Aya Nakae, Ryota Kanai & Hiroshi Ishiguro (2013) "Huggable communication medium decreases cortisol levels" *Scientific Reports* 3, Article number: 3034 doi:10.1038/srep03034 23 October 2013

3 Http://kidshealth.org/ (1995-2015) "Encouraging Your Child's Sense of Humor". Reviewed by D'Arcy Lyness, PhD (2012)

4 FAQ The Baby Laughter Http://babylaughter.net/more-information-2project

5 Provine, Robert R. (2001) *Laughter: A Scientific Investigation*. New York: Penguin

6 Http://kidshealth.org/ (1995-2015) "Encouraging Your Child's Sense of Humor". Reviewed by D'Arcy Lyness, PhD (2012)

7 Huckshorn, Kevin Ann. (2004) *Creating violence free and coercian free mental health treatment.* See Sensory Room. www.oregon.gov/DHHS/addiction/trauma-policy/reducing-use.ppt

   ROCKING CHAIR: The rocker has been shown to aid a 75% dramatic improvement in stress management at Cooley-Dickenson Hospital, Northhampton, MA. This figure represents 75% less use of restraint or seclusion.

8 Leandro Z. Agudelo, Teresa Femenía, Maria Lindskog et al.(2014) "Skeletal Muscle PGC-1α1 Modulates Kynurenine Metabolism and Mediates Resilience to Stress-Induced Depression." *Cell.*Volume 159, Issue 1, p33–45, 25 September 2014 http://www.cell.com/cell/abstract/S0092-8674%2814%2901049-6

8. Moe Moe Pepe

1 Vickhoff, Bjorn. et al. (2013) "Music structure determines heart rate variability of singers." *Frontiers in Psychology.* 09 July 2013. www. Frontiers in Auditory Cognitive Neuroscience.org.

9. Sleepy Little Baby

1 In 2004, I was invited to present Harp Massage Experiences at the Emanuel Hospital Eating Disorders Clinic with the staff. As the doctors and the nurses experienced the vibrational effect on their bodies, they noted increased warmth as the vibrations improved their general circulation. Daddy's voice has a similar effect. -LCS

10. Bis m'Allah

1 D. J. Shusterman, K. Jansen, E. M. Weaver and J. Q. Koenig (2007) "Documentation of the nasal nitric oxide response to humming: methods evaluation." *European Journal of Clinical Investigation* (2007) 37, 746–752 DOI: 10.1111/j.1365-2362.2007.01845.x note: The humming sample measured results from ten seconds duration and 3-5 intervals.

11. Safe in the Night 'A Rusty Old Halo'

1 Carter, Sue. (2003) "Developmental consequences of oxytocin." *Physiology & Behavior.* 79(2003) 383-397.

2 Carter, Sue. (2008) *Social bonds.* Portland State University. Also Gauvtik, et al. *PNAS*, 1996

3 Carter, Sue. (2003) "The developmental consequences of oxytocin." *Physiology & Behavior.* 79(2003) 383-397.

4 (Ibid.)

5 Porges, Stephen. (2007) "Brainstem nuclei that are oxytocin sensitive." *Love or Trauma?* DVD

6 (Ibid)

7 Carter, S. (2008) *Love or Trauma?: Social bonds.* 1 day Portland State University.

8 Carter, C. Sue (2019) Love As Embodied Medicine, *The Art and Science of Somatic Praxis.* Vol. 18, No.1 Spring 2019, pp.19-25

9 Carter, C. Sue (2021) The Healing Power of Love: An Oxytocin Hypothesis. Kinsey Institute. Indiana Univ. Delivered at the Solving Chronic Pain Summit. Jan.22-24. Open Center. NY.

10 Carter, C.Sue (2021) The Healing Power of Love: An Oxytocin Hypothesis. Kinsey Institue. Indiana Univ. Delivered at *Solving Chronic Pain Summit* Jan.22-24. Open Center, NY

13. Swedish Lullaby

1 Russo, MA., Santarelli, DM., O'Rouke, D. (2017) "The physiological effects of slow breathing in the healthy human." *Breathe*, 2017, 13.. 298-309

2 Eckberg, Dwain L. (2003) "The Human respiratory gate." *J. of Physiology.* Apr.15; 548 (Pt.2): 339-332 online 2003 Mar.7. Doi 10.1113/jphysiol.203.037192

3 Dawson, K.W. & Fischer, G. (1994) *Human Behavior and the Developing Brain.* New York: Guildford Press, p.9.

4 Mate, Gabor. (2010) *In the Realm of Hungry Ghosts.* Berkeley: North Atlantic Books. 191.

5 Carter, C.Sue (2021) The Healing Power of Love: An Oxytocin Hypothesis. Kinsey Institue. Indiana Univ. Delivered at *Solving Chronic Pain Summit* Jan.22-24. Open Center, NY

14. Mummy Tummy Yum

1 Beeman, Mark. Jung-, Bowden, Edward M et al.(2004) "Neural Activity When People Solve Verbal Problems with Insight." *PLOS* Published: April 13, 2004. DOI: 10.1371/journal.pbio.0020097

2 Sousa, David A. (2006) *How the Brain Learns.* Thousand Oaks: Corwin Press. p.231.

15. Yes, Jesus Loves " _____ " (Your baby's name here)

1 Fancourt, Daisy et al. (2016) "Singing modulates mood, stress, cortisol, cytokine and neuropeptide activity in cancer patients and carers." *E Cancer Journal.* Ecancer.org/journal/10

2 Vickhoff, Bjorn. et al. (2013) "Music structure determines heart rate variability of singers." *Frontiers in Psychology.* 09 July 2013. www. Frontiers in Auditory Cognitive Neuroscience.org

3 Keeler, Jason R. et al. (2015) "The neurochemistry and social flow of singing: bonding and oxytocin". *Front. Hum. Neurosci.* 23 September 2015/ https://doi.org./10.3389/fnhum.2015.00518

4 Carter, C.Sue (2021) The Healing Power of Love: An Oxytocin Hypothesis. Kinsey Institue. Indiana Univ. Delivered at *Solving Chronic Pain Summit* Jan.22-24. Open Center, NY

16. My Grandmother's Eyes (Joseph Dearest, Joseph Mild)

1 Becket Ebitz, R. and Platt, M. (2013). *Frontiers of Behavioral Neuroscience.*" An evolutionary perspective on the behavioral consequences of exogenous oxytocin application" Front Behav Neurosci. 2013; 7: 225. Published online Jan 17, 2014. doi: 10.3389/fnbeh.2013.00225 figure 1

2 Dias, Brian G and Ressler, Kerry. (2013) "Parental olfactory experiences influences behavior and neural structure in subsequent generations." *Nature Neuroscience* 17. 89-86 (epigenetics mouse)

3 Kuhn, Merrily. (2013) "Understanding the Gut Brain: Stress, Appetite, Digestion & Mood." www.ibpceu.com

4 Mate, Gabor. (2010) *In the Realm of Hungry Ghosts.* Berkeley: North Atlantic

5 Kuhn. Ibid.

6 Mate. 198-200

7 Kuhn. Ibid.

8 Kang, Scholp and Jiang (2017) A review of the physiological effects and mechanisms of singing. Elsevier Inc Pub. For *The Voice Foundation.*

17. Loo La La Loo

1 McGilchrist, Iain (2012-07-15). *The Divided Brain and the Search for Meaning* (Kindle Locations 262-266). Yale University Press. Kindle Edition.

2 Hatfield E, Cacioppo JT, Rapson RL. "Emotional contagion." *Curr Dir Psychol Sci* 1993;2:96–99.

18. Rock A Baby Bye - Putting Family Names in the Song

1 Porges, Stephen. (2011) *The Polyvagal Theory: Neurophysical Foundations of Emotions, Attachment, Communication, Self-Regulation.* New York: Norton. pg.112

The analysis revealed that there was a strong correlation between the baby's use of the Social Circuit pathway, and their three year-old positive interactions with others, and their appropriate social behavior.

"... tasks that require social interaction and/or focused attention to external stimulus should be and are in this study, related to fewer social problems (lower scores on the social withdrawal and aggressive scales, lower scores on the depressed scale.)" Porges (2011)

The babies with high baseline vagal tone factors were happier, more social, slept better.

2 Ibid. p.145.

3 Ibid. 112.

19. Sri Ram Jai Ram

1 Hatfield, E. Bensman, L. Thorton, P. and Rapson, R. (2014) "New Perspectives on Emotional Contagion: A Review of Classic and Recent Research on Facial Mimicry and Contagion." Univ. of Hawaii: *Interpersona.* Vol.8 no.2.

21. Frere Jacques

1 Provine, Robert R. (2000) *Laughter: A Scientific Investigation.* New York: Penguin

2 http://babylaughter.net/more-information-2 FAQ/The Baby Laughter project

3 Ibid.

4 http://www.pbs.org/thisemotionallife/topic/humor/benefits-humor

5 Ibid

6 Ibid

22. Shona Lullaby

Re: SINGING MAKES MUSCLES MORE EFFICIENT

1 Thomas Hans Fritz, Hardikar, S. et al. (2013) "Musical agency reduces perceived exertion during strenuous physical performance." *PNAS.* vol. 110 no. 44 17784–17789, doi: 10.1073/pnas.1217252110

Re: MUSCLES MAKE NATURAL ANTIDEPRESSANT

2 Femenia, T., Orhan, F. et al. (2014) "Skeletal Muscle PGC-1α1 Modulates Kynurenine Metabolism and Mediates Resilience to Stress-Induced Depression." *Cell.*Volume 159, Issue 1, p33–45, 25 September 2014. http://www.cell.com/cell/abstract/S0092-8674%2814%2901049-6

Results: Exercise that challenges the muscles can neutralize a stress-induced toxin called kynurenine which also triggers depression symptoms. Challenging exercise produces PGC-1 alpha-1, which stops kynurenine from reaching the brain. The brain is protected.

Re: LANGUAGE-LEARNING WINDOW

3 Kuhl PK. (2007) "Is speech learning 'gated' by the social brain? *Dev Sci.* 2007 Jan; 10(1):110-20.

4 Kuhl, Patricia K. (2010) "Brain Mechanisms in Early Language Acquisition." *Neuron.* 2010 Sep 9; 67(5): 713–727. doi: <u>10.1016/j.neuron.2010.08.038</u> PMCID: PMC2947444 NIHMSID: NIHMS2343566

5 Kuhl. Ibid.

6. Grape C., Sandgren M., Hasson, L.O. et al. (2002) Does singing promote well-being? An empirical study of professional and amateur singers during a singing lesson. *Integr. Physiol. Behav. Sci.* 2002; 38: 65-74.

7 Carter, C. Sue. (2019) Love As Embodied Medicine. International Body Psychotherapy Journal *The Art and Science of Somatic Praxis.* Vol.18, Number 1,Spring 2019, pp.19-25.

8. Carter. Ibid

23. "Poppy, Sing the Nap Song."

1 Siegel, Daniel. (2007) *The Mindful Brain.* New York: W.W. Norton & Co. p.104 fl

25. Charlotte's Lullaby

1 <u>https://developingchild.harvard.edu/science/key-concepts/brain-architecture/#neuron-footnote</u> 700 new synapses (neural connections) are formed every second in the first year of life (Shonkoff). In 2017, the figure was changed to over a million.

2 DeCasperA, & Cartens, A. (1981). "Contingencies of stimulation: Effects on learning and emotions in neonates." *Infant Behavior and Development*, 4, 10-35.

3 Csikszentmihalyi, Mihaly. (1990) *Flow: The Psychology of Optimal Experience.* New York: HarperCollins.

4 Kreutz;

and

Keeler, Jason R. et al (2015) "The neurochemistry and social flow of singing: bonding and oxytocin" *Front. Hum. Neurosci.,* 23 September 2015 | <u>https://doi.org/10.3389/fnhum.2015.00518</u>

5 Kang, Scholp and Jiang (2017) A review of the physiological effects and mechanisms of singing. Elsevier Inc Pub. For *The Voice Foundation.*

6 Ibid.

7 White, E. B. (1952) *Charlotte's Web.* New York: Harper & Bros. P.104. The children heard the words that follow:

> *"Sleep, sleep, my love, my only*
> *Deep, deep, in the dung and the dark.*
> *Be not afraid and be not lonely!*
> *This is the hour when frogs and thrushes*
> *Praise the world from the woods and the rushes."*

Because of the 30 minute time constraint, we used an ABA[1] music form with the words as well, using the first line again. Decades later, the lyrics had drifted and morphed in my memory, yet the melody stayed constant. The rendition is sung in Wind Painting technique.

26. The Night Night Song

Re: PHYSIOLOGICAL CHANGES FROM SINGING

1 Kreutz, G. and Bongard S. et al (2004) "Effects of choir singing or listening on secretory immunoglobulin A, cortisol, and emotional state." *J Behav Med.* 2004 Dec;27(6):623-35. https://www.ncbi.nlm.nih.gov/pubmed/15669447

1 & 2 Schladt, T. Moritz et al. (2017) "Choir versus Solo Singing: Effects on Mood, and Salivary Oxytocin and Cortisol Concentrations" Front Hum Neurosci. 2017; 11: 430. Published online 2017 Sep 14. doi: [10.3389/fnhum.2017.00430]PMCID: PMC5603757PMID: 28959197

3 Fancourt, Daisy. And Williamon, A. et al. (2016)"Singing modulates mood, stress, cortisol, cytokine and neuropeptide activity in cancer patients and carers" Ecancermedicalscience. 2016; 10: 631. Published online 2016 Apr 5. doi:[10.3332/ecancer.2016.631]PMCID: PMC4854222 PMID: 27170831

3 Ibid

4 Ibid. Kreutz; and

Keeler, Jason R. et al (2015) "The neurochemistry and social flow of singing: bonding and oxytocin" *Front. Hum. Neurosci.,* 23 September 2015 | https://doi.org/10.3389/fnhum.2015.00518 1

5 Osman, Sara E.,Tischler, Victoria, **and** Schneider, Justine (2016) "Singing for the Brain': A qualitative *Dementia (London)*study exploring the health and well-being benefits of singing for people with dementia and their carers." 2016 Nov; 15(6): 1326–1339.Published online 2014 Nov 24. doi: [10.1177/1471301214556291] *PMCID*: PMC5089222 PMID: 25425445

Re: SINGING MAKES MUSCLES MORE EFFICIENT

6 Thomas Hans Fritz, Samyogita Hardikar,Matthias Demoucron[a], et al. (2013) "Musical agency reduces perceived exertion during strenuous physical performance." *PNAS.* vol. 110 no. 44 17784–17789, doi: 10.1073/pnas.1217252110

7 Ibid Kreutz, Ibid Schladt

8   Russo, MA., Santarelli, DM., O'Rouke, D. (2017) "The physiological effects of slow breathing in the healthy human." *Breathe,* 2017,   13.. 298-309

Re: PHYSIOLOGICAL CHANGES FROM INFANT CARE-GIVING

9 Feldman, Ruth. (2020) On the series *"Babies",* Documentary Series, episode #1. Love.

10 Carter, C.Sue (2021) The Healing Power of Love: An Oxytocin Hypothesis. Kinsey Institue. Indiana Univ. Delivered at *Solving Chronic Pain Summit* Jan.22-24. Open Center, NY

11 Ibid Carter (2021)

12/ Kang, Jing. Scholp and Jiang (2017) A review of the physiological effects and mechanisms of singing. Elsevier Inc Pub. For *The Voice Foundation.*

13 Ibid. Kang, Jing.

27. Silent Night

1 Siegel, Daniel J. (2012) *Pocket Guide to Interpersonal Neurobiology.* New York: WW Norton & Co. p. 27-2

28. Grandbaby's Lullaby "Sleep Well"

1 Panksepp, Jaak. (1998) *Affective Neuroscience: The Foundations of Human and Animal Emotions.* New York: Oxford Press

2 Panksepp, Jaak. & Biven, Lucy. (2012) *The Archeology of Mind: Neuroevolutionary Origins of Human Emotions.* New York: Norton & Co. p.308.

3 Ibid.

4 Ibid.

5 Ibid.

6 Ibid. p.363; and Panksepp, Jaak. (1998) *Affective Neuroscience: The Foundations of Human and Animal Emotions.* New York: Oxford Press

7 Ibid. Pankseep, *Archeology* p.367. & 304

8 Ibid. Panksepp. p.368.

9 Ibid

10 Ibid p.355.

11 Ibid p.367.

12 Ibid p.371

13 Ibid p.360

14 Ibid p.371

15 Ibid p.380

16 Ibid p.379

17 Ibid p.380

18 Ibid p.313 citing Hrdy, (2009) *Mothers and Others.* Cambridge, MA: Harvard Univ. Press.

and

18b Kosfeld, M., Heinrichs, M.et al. (2005) "Oxytocin increases trust in humans." *Nature* 2005 Jun 2;435(7042):673-6.

29. Hold You Close

1 Siegel, Daniel and Bryson, Tina Payne. (2011) *The Whole-Brain Child.* New York: Bantam.

2 Porges, Stephen. (2011) *The Polyvagal Theory: Neurophysical Foundations of Emotions, Attachment, Communication, Self-Regulation.* New York: W.W. Norton & Co.

3 Ibid

4 Ibid

31. Julian's Lullaby

1 Vickhoff, Bjorn. et al. (2013) "Music structure determines heart rate variability of singers." Frontiers in Psychology. 09 July 2013. www. Frontiers in Auditory Cognitive Neuroscience.org.

2 Gopnik, Alison. (2011) What do babies think? Talk Video. TED.com www.ted.com/talks/alison_gopnik_what_do_babies_think?

3 Park, BJ, Tsunetsugu, Y. et al. (2010) "The physiological effects of Shinrin-yoku (taking in forest atmosphere or forest bathing): evidence from field experiments in 24 forests across Japan." *Environ Health Prev. Med.* 2010 Jan 15(1).18-26.

4 Ibid.

5 Ibid.

6 Report by National Chung Hsing University Reveals the Secrets of Phytoncides, Campus News, Ministry of Education Republic of China (Taiwan) February, 7 2013. english.moe.gov.tw/ct.asp?xItem=10743&ctNode=11020&mp=1

7 Li Q et al. "Phytoncides, the essential oils in the bark and stems, induce human natural killer cell activity." *Immunopharmacol Immunotoxicol* Immunopharmacol Immunotoxicol. 2006;28(2):319-33. Source: Department of Hygiene and Public Health, Nippon Medical School, Tokyo, Japan. qing-li@nms.ac.jp

8 Wohlleben, Peter. (2015) *The Hidden Life of Trees.* Vancouver/Berkeley: Greystone Books; and

8a Jee-Yon Lee and Duk-Chul Lee, "Cardio and Pulmonary Benefits of Forest Walking versus City Walking in Elderly Women: A Randomized, Controlled, Open-Label Trial," *European Journal of Integrative Medicine* (2014): 5-11.

9 Ibid.

10 Taylor, A.F. & Kuo,F. (2008) "Children with attention-deficit concentrate better after walk in the park." *Journal of Attention Disorders.* 20 (10) April p.1-7.

11 Wohlleben, and Harmuth, Frank, et al. (2003) *Der sachsische Wald im Dienst der Allgemeinheit* ("The woods of Saxony in the service of the general public") State Forestry Commission of Saxony. www.smul.sachsen.de/sbs/download/Der_sachsische_Wald.pdf.

The leaves and needles catch particulates as they float by. Each year a square mile of diverse forest has been known to remove 20,000 tons of material: soot, pollen, dust, acids, toxic hydrocarbons, nitrogen compounds, and also reduces air-borne germs.

12 Kawakami K. et . (2004) "Effects of phytoncides on blood pressure under restraint stress in SHRSP." *Clin.Exp.Pharmacol.Physiol.* 2004 Dec.31 Suppl.2:S27-8.

13 Kimmerer, Robin Wall. (2013) *Braiding Sweetgrass*. Minneapolis: Milkweed Publishers. pg. 236

32. Compassion Lullaby (May Our Hearts Enfold You)

1 Adam D. I. Kramer, Jamie E. Guillory, and Jeffrey T. Hancock.(2014)"Experimental evidence of massive-scale emotional contagion through social networks." *PNAS* June 17, 2014. 111 (24) 8788-8790. https://doi.org/10.1073/pnas.1320040111

2 Mate, Gabor. (2010) *In the Realm of the Hungry Ghosts*. p.162

3 Ibid. 151

4 Ibid. 171.

5 Ibid.161.

6 Ibid.162.

Part II: Simple Soothers

44. Calming the Fussy Hour

1 Kuhn, Simone et al. (2017) "In search of features that constitute an "enriched environment" in humans: Associations between geographical properties and brain structure" *Scientific Reports* 7, Article number: 11920 (2017)

2 Moss M, Oliver L. (2012) "Plasma 1,8-cineole correlates with cognitive performance following exposure to rosemary essential oil aroma." Ther Adv Psychopharmacol. 2012 Jun;2(3):103-13. Doi: 10.1177/2045125312436573.

3 Atsumi T. and Tonosaki K. (2007) "Smelling lavender and rosemary increases free radical scavenging activity and decreases cortisol level in saliva." *Psychiatry Res* (2007) Feb 28;150(1):89-96. Epub 2007 Feb 7; and

3a Kandhalu, Preethi. (2013) "Effects of cortisol on physical and psychological aspects of the body and effective ways by which one can reduce stress." *Berkeley Scientific Journal* • Stress • Fall 2013 Volume 18 • Issue 1, 14-16.

4 Park, BJ, Tsunetsugu, Y. et al. (2010) "The physiological effects of Shinrin-yoku (taking in forest atmosphere or forest bathing): evidence from field experiments in 24 forests across Japan." *Environ Health Prev. Med.* 2010 Jan 15(1).18-26.

5 Cheng, W., Lin, C., Chu, F. *et al.* Neuropharmacological activities of phytoncide released from Cryptomeria Japonica. *J Wood Sci* **55,** 27–31 (2009). https://doi.org/10.1007/s10086-008-0984-2

48. Remember to drink water...

1 Sousa, David A. (2006) *How the Brain Learns.* Thousand Oaks: Corwin Press.

Glossary References:

Paaksepp, Jaak. (2012) *The Archeology of the Mind.* NY: W.W. Norton & Co

Paaksepp, Jaak. (1998) Affective Neuroscience: The Foundations of Human and Animal Emotions. NY: Oxford Univ. Press.

Porges, Stephen. (2004) Neuroception: A subconscious system for detecting threats and safety. *Zero to Three.* May, 19-24.

Porges, Stephen. (2011) *The Polyvagal Theory: Neurophysical Foundations of Emotions, Attachment, Communication, Self-Regulation.* New York: Norton.

Siegel, Daniel J. (2012) *Pocket Guide to InterPersonal NeuroBiology.* NY: W.W. Norton & Co

Siegel, Daniel J. and Bryson, Tina Payne. (2012) *The Whole-brain Child.* NY: Bantam Books Trade Paperbacks.

Glossary by Baby

The neuroscience and circuitry within these pages is a partial picture of the complex and highly sophisticated processes of the body-brain-mind that we each inhabit. This information opens us to the fact that each touch, each voice sound, each motion, each interaction, each look or absence of eye contact, changes the baby's brain. We can be a gift to each others' genetic expression.

affective neuroscience – With pioneering research by Jaak Panksepp and others, researchers have identified seven primary affective states in the mammalian brain. (This area is sometimes called the mid-brain or limbic brain.) Each is hard-wired and defined by brain regions, neurotransmitters, neurochemicals, and automatic, consistent neurocircuits. They are CARE (nurturance), SEEKING (expectancy), PANIC/GRIEF (separation distress in the infant), FEAR (anxiety), PLAY (social joy), RAGE (anger) and LUST (sexual excitement). It's about emotions!

The following definitions are worded from the baby's point of view.

attachment – This is my relationship with you. *Attachment* is a term for my development of emotional responses and neurosynaptic pathways in my baby brain, in response to your signals and behaviors. There are several types:

> **secure attachment** – My needs are met. You really see me. I feel seen. You enjoy me and I grow optimally with curiosity and kindness. I branch out, learning to enjoy others. I sleep well. I learn to self-soothe from *you soothing me* and providing safety. I explore with full attention. My world makes sense. It is secure. Yours is 'good enough' parenting. I am generally happy.

> **insecure-anxious** – With inconsistency my needs are met. I am clingy and fearful. I am fretful about life and fretful near others.

> **insecure-avoidant/ambivalent** – I may try to avoid needing you and may resist you picking me up. I may be controlling...or...I may be a bottomless pit of neediness.

> **insecure-disorganized** – I am threatened by you, but I need you. I can't adapt. What a mess.

attunement – When you focus your attention on my internal world, I feel connected to you, I feel 'felt'. From this safety, I integrate my neural information; I get to know you and get to know me. Attunement is the alignment of verbal and non-verbal cues between the two of us. We flow.

bounce-walking – I may like a gentle bounce as you carry me. It gives the sensation of regular, safe motion in my body. You can delicately bounce with your arms, knees, or on the balls of your feet. Make sure to support my head. It relaxes my brain and benefits my circulation and lymph system for better health.

contingent communication – I yawned. My eyelids were lowering. The big person noticed, picked me up and said, "Let's get you ready for naptime."

emotional contagion – I have an unconscious tendency to mimic and synchronize to your facial expressions, vocalizations, postures, movements and emotions. It's our human phenomenon: it's catching!

endogenous opioids – I feel good! Na-na-na-na-na-na-nah. Just the best that I could. My central nervous system and pituitary gland make these neurotransmitters in me. These experiences create such intense rewards between us.

gaze – I love to gaze at your face. In a staring contest, I win. Just give me calm friendliness while I stare into your eyes. You're wonderful!

"good enough" parenting – You attend to most of my needs most of the time. If we are tumbling out of alignment (too busy, errands make us miss the nap time, long car rides, no time for play, hurrying all the time) we enter a cycle of 'rip and repair'. We feel disconnected – then long to reunite. We reconnect – and repair – then plan for better and enter back into humming along. I am a wise baby and I know that no parent is perfect. Just teach me the art of repairing relationships. Together, we can tune to each other and our needs. I am learning from you.

implicit memory – I have memories, sensations, repeated experiences, emotions, behavior patterns and perceptions that are beyond words or any sense of an event. My **explicit memory** will come online around when I am 15-18 months old. Then I will be able to remember events and words.

integration – The ability of my synapses to communicate and gain information from all the brain's neuropathways and stored memories. I am building 90% of my brain after I am born, most of it in the next three years. Whew, that's miles of circuits. I think I need a nap.

interoception – J.P. in #14 has a life-time of family and professional experience with the healthy, the needy, and people in transition in their lives. She has social ethics integrated with her skills of reading people, their social cues and values. She can process the character information in 2 seconds. Her self-knowledge i.e. *interoception,* informed her in her decision to house the young refugee.

Interpersonal Neurobiology – This is how I grow my body-brain-mind. You are my everything! The wisdom from the science of Interpersonal Neurobiology (IPNB) is that we physically change, and *do* change our neuro-pathways as we interact with each other. See also mindGAINS.org

invariant representation – I see an owl on a bib. I see an owl on an alphabet sketch. I see a photo of an owl on a calendar. And the next owl representation that I see regardless of color or detail, prompts me to say "hoo hoo." I have an area in my neo-cortex just for this. It's updating its data base continually.

laughter – This is just the best for my healthy body. I may laugh over 300 times a day when I'm a toddler. My hilarity releases "feel-good' endorphins in my body.

mindfulness – The ability to bring my physical body and my thoughts to the same place at the same time! As a baby, I have a beginner's mind!

mindsight – My mom senses how my mind is thinking. She also knows her own mind.

mirror neurons – If you stick out your tongue at me; even if I am hours old, I 'll stick out my tongue to you in imitation.

"motherese" – Anne Fernald PhD, Stanford University coined this word to describe an internationally prevalent style of rising and falling prosody that mothers all over the world instinctively do with their babies. She noted four types. You may do it without thinking, and I love it! It is like gentle touch to me. Currently some researchers are referring to it as **"parentese"**.

myelination – As I learn, I form nerve branches called 'dendrites.' Myelin is an insulating sheath I will form around my new dendrites. Myelin allows them to fire with much greater speed and no interfering crosstalk. Until then, I do many things slowly, like responding to you. As a baby/toddler, it could take me 16 seconds to follow a command: to hear it, to figure out what you said, to find the parts of my body that need to respond, to start my body in motion, and then execute the action with a smile. When I'm older, you'll see how fast I move! Please be patient. I'm a work in progress.

"parentese" – This is a universal linguistic style that adults use to talk to me. It is often in slower, simpler, repetitive words, with more exaggerated sounds. It can be higher pitched than regular talk, like the sound of my higher infant/toddler voice. It says to me that you are really totally with me. If you want, you can learn it online! (see also **"motherese"**)

parietal cortex – My parietal cortex is at the top of my brain. As I build it, I will begin to know what is me and where I am in space. I have no idea when I am born. I don't even know that you and I are separate. I will also map where you are for my safety.

phytoncides – These are chemical essences made by trees. When I enter a forest, I breathe in the forest-filtered, pure, clean air. Within 20 minutes the trees relax me, help me to sleep better, breathe easier, and lower my blood pressure and stress hormones. They also help me to pay better attention. I like trees. My brain likes their fractal patterns and green hues.

resiliency – When I go through something painful or traumatic, little or big, I am able to repair and get on with life. I get more resilient when I have safety, face time with you, and play with which to build my nerve pathways to my Pre-Frontal Cortex.

respiratory gating – This is my ability to use my breath to calm myself with slow breathing i.e. 4-10 breaths per minute, or singing, which often lets me breathe at that rate, especially for lullabies. You can use it, too.

serotonin – This is my 'feel good' neurotransmitter. I make 95% of it in my abdominal intestines, and the other 5% in my brain.

synapse – This is the tiny gap between my neurons. Impulses pass across it with the diffusion of a neurotransmitter. Then my nerves are firing. Each of my nerves can form up to 10,000 synapses connecting to other nerve cells! And I will be growing

GLOSSARY

80 to 100 billion nerve cells with your help. I am adapting to your world for the best living we can get with my body!

time-in – Any practice of reflecting inwardly on Sensations – Images – Feelings – Thoughts = SIFT. This is a mindfulness activity and brings neural integration. I love it when you do this daily!

track – I use all my sensations, senses and close attention to focus on you. It's all new to me. Watching you helps me to wire together my own self- awareness. This stimulates neuronal activation and growth. (Seigel POCKET AI-74) I know that my 'neurons that fire together, wire together'. (Carla Shatz describing Hebb's Law 9-2) I track to find out what I am, where I am and what is me. It's all new.... this "me."

vagus nerve (X) – It is just the most important nerve in my baby body-brain running from my brain stem and facial muscles, to my heart, lungs and digestive system. My auditory nerve connects to it directly. That's why a gentle voice helps to soothe me.

vagus nerve – dorsal branch - If I have fainted, go into shock, or become immobilized in life-threatening circumstances, then I am using this oldest unmyelinated branch of the Vagus Nerve to protect my vital organs and life.

ventral vagus nerve – When I am calm, social, making eye contact, and building my brain-body for optimum living, I am using this newer ventral Social Engagement Circuit of the Vagus Nerve.

wind painting – Sound that combines the "shhh" sound from the tongue and alveolar ridge with changes in the mouth and throat to produce pitches. The breath is like blowing to cool a bite of hot mashed potatoes in your mouth. It soothes me.

For Further Reading

Emoto, Masaru. (2004) *Love Thyself: The Message From Water vol.3.* Toyko: Kazuko Emoto.

Gopnik, Allison. (2016) *The Gardener and the Carpenter: What the New Science of Child Development Tells Us About the Relationship Between Parents and Children.*
New York: Farrar, Straus & Giroux.
Also, (1999) *How Babies Think*
and (1999) *The Scientist in the Crib.*

Karp, Harvey. (2002) *The Happiest Baby on the Block* book, DVD, audio

Maman, Fabien. (1997) *Book III The Body as a Harp: Sound and Acupuncture.*
Redondo Beach: Tama-Do Press,.   .

Siegel, Daniel and Bryson, Tina Payne. (2011) *The Whole-Brain Child.* New York: Bantam.

DVDs

(2000) *Brighter Baby,* hosted by Brenda Adderly. New York:Wellspring Media
ISBN 1-58350-213-0 Brenda has great technique for baby and toddler massage.

Moore, Diana. (2004) *Baby's First Touch.* Loving Touch Foundation.
ISNB 1-883606-02-0
Used this as well to awaken a grandchild's sense of joy, and calmness at naptime.

APPENDIXES

#31. Julian's Lullaby

Licia Claire Seaman

* Your child's name here.

This song was created for a hospitalized child. Replace "Julian" with your own child's name.

Song Key Signature Distribution

All of the songs are rendered to staff notation in the keys of the singers' originally recording. Below is the data distribution. It seems that any pitch can be just right for the moment. Feel free to pitch the songs where your voice is comfortable. Pitch distribution is a function of the brain's rostromedial prefrontal cortec (rmPFC) which distributes the tones in clusters that musicians would recognize as related scale tones.

key	C	C/Db	D	Eb	E	F	F#/Gb	G	Ab	A	Bb	B
Major	7	2	1		1	6	1	1	1	3	1	4
minor					2	1				3		1

key	C	C/Db	D	Eb	E	F	F#/Gb	G	G#/Ab	A	Bb	B
Major	#1 #5 #10 #12 #19 #21 #31	#18 #26	#24		#7	#4 #8 #9 #15 #25 #27	#14	#11	#3	#6 #29 #34	#2	#16 #22 #28 #32
minor					#13 #23	#30				#20 #40 #42		#17

Janata, Petr. And Birk,J. et al (2002) "The cortical topography of tonal structures underlying Western music." *Science* 2002 Dec 13:298 (5601): 2167-70. DOI: 10.1126/science.1076262

Lullabies – Interview Form

Name_____ Date_____ Ph#_____

Contact info:_____

Do/did you sing to your kids?

Do you sing to your grandkids?

Did someone sing to you?

Names of Songs _____

_____

Associated with: Naps____ Being rocked___ Cuddling___ Carried___ Bedtime____

What did it look like when you were singing/sung to?

What room did you use?_____ Outside?____ Car?____ Occasion?_____

Time of Day_____ Frequency_____ Any routine?_____

Duration: 1-5 minutes   5-10 minutes   10-20 minutes+

Social (Others sing, too?)_____ Child sings with you?_____

Physical Interaction: sitting side-by-side   child in bed   adult lying on child's bed   in a rocker

   child on lap   at a piano   carrying child   dancing/swaying together   child active

Did you ever hum?   Whisper?   Make up your own?   phrase? vocables? words?

Language used? Eng   Other_____

Result of Lullaby: awake   laughing   jumping on the bed   dancing   playing   sleepy   asleep

   cooing   smiling   quiet-alert   comforted   stopped crying   child singing   agitated

Record_____ Labeled _____ Length_____ Sheet Music___ Words___Purchase___

Permission to use original song?_____

Permission to quote_____ This is being recorded for a field recording for the typist to transcribe.

Permission to use voice recording_____ There is also a chance that we might show people how many different kinds of voices can sing a lullaby. So that they might say, "I can do this! I can lull."

Questions of me?_____ How does it feel after memories?_____

   Name an emotion? _____ Where in your body do you feel it?_____

Acknowledgements

Dietrich Gruen Seaman – assistant editor, cover art design, audio editing, formatting, and tech man, ...*and for enduring all his sleepy yawns while editing the lullaby singing tracks.*

Joshua Gruen Seaman – Spanish & French translation assistance, photos.

Karen Gatens – layout designs

Mel Zimmerman – Baby icon & Child icon

Sarah Van Roo – proof-reading and inspiration

Mark Baumann – "Loo La La Loo" research quotes. Thanks, Mark.

Debra Pearce-McCall, PhD – the science analysis on "Baby's Singing Talks to Me"

Erica Azim – provided the written Shona on "Auwe Nyrara, Mwana".

Lorain Fox Davis – Advisor, for *Aztec Lullaby* pronunciation & spelling

Additional Interviewers – Anne Caruso, Julie Granger, Joan Bouwles, Karen Gatens

Bev Gay – photo support

Christina & Monroe Gay – #2 Illustration models

Manamaya Rayanalynn Peterson – #7 Illustration model

Grandmother Susan & Baby Skye – Cover Illustration models

Demetria Ford – #7 concept drawing

Amy McCandlish Esper – photo support

Bernie & Kathy Casey – #34 "ma chuisle"

Readers – Jackie Loomis, Ann Di Loreto, Frosti Talley, Catherine Bax, Julia Hejja.

Jane Braunger, Richard Corbett

Los Lobos Writers Retreat 2018, Chile

And to that special friend who helped this book to come alive.

Beta testers "Muffin", "Junebug", and "Bebsi".

Also, I thank the many people who gave their spirits in enthusiastic stories and serene songs by the hours. I am grateful to have their openness and willingness to share their voices and songs for the many in generations to come. And their mates' assistance for the ones journeying down the path of dementia. Thank you, those of you who called them back to such happy memories.

Photo credits

(Reference photos for illustrator, except * for actual images used and/or composited)

Joshua Seaman Photography (www.relicpro.com): #4, #10, #28

Dietrich Seaman: front cover, #6, #7, #16, #18, #21, #23, #31, rear cover*

Jen Bolton: #10 "Bijou & Poppy", #32 "Bebel & Dad"

Erica Azim: #22

Gordon Gay: #2

Delila Woodruff: #19

Chris Lum: #27

The Fry family: #24

The Hicks family: #15

The Fleming extended family: #3 "Che-Che & Mima", #9 "El-Bell & Papi"

The Lippold family: #17

The Villaume family: #30

The Talla family: #14 "Zacharie Talla & Grandmother"

Shutterstock.com (standard license):
    #8 (1218839440), (1269798856)
    #12 (708518833)
    #13* (1710766792)
    #20 (1554631832), (1790405243)
    #22 (1755340601)
    #25* (13686619)
    #26* (1686811894), (101619499), (1162758040), (1649094598), (43430971),
        (1311842888), (109053074), (115909030), (1331237546), (765510337),
        (292955921)
    #29 (737227969)

Dreamstime.com (licensed): Part I & Part II borders (46003905)

iStockphoto.com (standard license): #5 (172736473) "Old Hands, Young Hands"

Clipart-library.com: Introduction, world map outline (681720)

ACKNOWLEDGEMENTS

References

Adam, D. I. Kramer, Guillory,J. and Hancock, J. (2014)"Experimental evidence of massive-scale emotional contagion through social networks." *PNAS* June 17, 2014. 111 (24) 8788-8790. https://doi.org/10.1073/pnas.1320040111

Addyman, Caspar & Addyman, I. (2013) "The science of baby laughter" *Comedy Studies*, 4:2, 143-153, DOI:10.1386/cost.4.2.143_1

Ainsworth, M., Blehar, M., Waters, E. et al.(1978) *Patterns of attachment: A psychological study of the strange situation.* New York: Psychology Press.

Agudelo, Leandro, Femenia, T., Lindskog, M. et al. (2014) "Skeletal Muscle PGC-1α1 Modulates Kynurenine Metabolism and Mediates Resilience to Stress-Induced Depression." *Cell.*Volume 159, Issue 1, p33–45, 25 September 2014 http://www.cell.com/cell/abstract/S0092-8674%2814%2901049-6

Atsumi T. and Tonosaki K. (2007) "Smelling lavender and rosemary increases free radical scavenging activity and decreases cortisol level in saliva." *Psychiatry Res* (2007) Feb 28;150(1):89-96.

Becket Ebitz, R. and Platt, M. (2013). " An evolutionary perspective on the behavioral consequences of exogenous oxytocin application" *Frontiers of Behavioral Neuroscience.* Front Behav Neurosci. 2013; 7: 225. Published online Jan 17, 2014. doi: 10.3389/fnbeh.2013.00225

Beeman, Mark. Jung-Bowden, Edward M. et al.(2004) "Neural Activity When People Solve Verbal Problems with Insight." *PLOS* Published: April 13, 2004. DOI: 10.1371/journal.pbio.0020097

Blackwell, L. A., Trzeniewski, K. H., & Dweck, C. S. (2007). "Theories of intelligence and achievement across the junior high school transition: A longitudinal study and an intervention." *Child Development*, 78, 246-263

Blackwell, Lisa. *Brainology Manual* p.17. Mindset Works Inc, mindsetworks.com. (2002-2013)

Bond, Allison. (2014) "Sing Your Way to Fitness" *Sci Mind* May/June 2014, p.17.

Carter, C. Sue. (2021) The Healing Power of Love: An Oxytocin Hypothesis. Kinsey Institue. Indiana Univ. Delivered at *Solving Chronic Pain Summit* Jan.22-24. Open Center, NY

Carter, Sue. (2008) Love or Trauma: Social bonds. Portland State University

Carter, Sue. (2003) Developmental consequences of oxytocin. *Physiology & Behavior.* 79(2003) 383-397.

Carter, Sue., Porges, Stephen. (2013) "The biochemistry of love: an oxytocin hypothesis." EMBO Rep. 2013 Jan; 14(1): 12–16 Published online 2012 Nov 27. doi:[10.1038/embor.2012.191] https://www.ncbi.nlm.nih.gov/pmc/articles/PMC3537144/#b7

Cheng, W., Lin, C., Chu, F. et al. (2009) Neuropharmacological activities of phytocide released from *Cryptomeria japonica. J Wood Sci* **55,** 27–31 (2009). https://doi.org/10.1007/s10086-008-0984-2

Cozolino, Louis. (2006) *The Neuroscience of Human Relationships.* New York: Norton.

Crittenden, Patricia McKinsey; Landini, Andrea (2011-06-13). *Assessing Adult Attachment: A Dynamic-Maturational Approach to Discourse Analysis.* New York: Norton, p. 186.

Crowley, Chris & Lodge, Henry s. (2004) *Younger Next Year.* New York: Workman Pub.

Csikszentmihalyi, Mihaly. (1990) *Flow: The Psychology of Optimal Experience.* New York: HarperCollins.

Dawson, K.W. & Fischer, G. (1994) *Human Behavior & the Developing Brain.* New York: Guildford Press

DeCasperA, & Cartens, A. (1981). Contingencies of stimulation: Effects on learning and emotions in neonates. *Infant Behavior and Development*, 4, 10-35.

Diamond, Marion Cleeves. (1988) *Enriching Heredity.* New York: The Free Press/Simon and Schuster.

Dias, Brian G and Ressler, Kerry. (2013) "Parental olfactory experiences influences behavior and neural structure in subsequent generations." *Nature Neuroscience* 17. 89-86 (epigenetics mouse)

Dykema, Ravi. (2006) Don't talk to me now, I'm scanning for danger: How your nervous system sabotages your ability to relate. *NEXUS*. March-April 2006. 30-35.

Eckberg, Dwain L. (2003) "The Human respiratory gate." *J. of Physiology*. Apr.15; 548 (Pt.2): 339- 332 online 2003 Mar.7. Doi 10.1113/jphysiol.203.037192

Fancourt, Daisy et al. (2016) Singing modulates mood, stress, cortisol, cytokine and neuropeptide activity in cancer patients and carers. *E cancer medical science*.2016, 10:631. E cancer.org/journal/10

Farroni, Csibra,Simion & Johnson (2002) Eye contact detection in humans from birth. *Proceedings of the National Academy of Sciences.* 99.9602-9605.

Feldman, Ruth. (2020) On the series *"Babies"*, Documentary Series, episode #1. Love.

Femenia, T., Orhan, F. et al. (2014) "Skeletal Muscle PGC-1α1 Modulates Kynurenine Metabolism and Mediates Resilience to Stress-Induced Depression." *Cell*.Volume 159, Issue 1, p33–45, 25 September 2014. http://www.cell.com/cell/abstract/S0092-8674%2814%2901049-6

Fernald, Anne. (2007) WNYC's Radio Lab with Jad Abumrad and Robert Krulwich delving into the making of their acclaimed program that melds science, philosophy, and psychology this podcast "short" from November, (2007),episode "Sound as Touch" with Dr. Anne Fernald, Director, Center For Infant Studies at Stanford.

Fields, R.. Douglas. (2009) *The Other Brain*. New York: Simon & Schuster.

Fritz, Thomas et al. (2013) "Musical agency reduces perceived exertion during strenuous physical performance", *PNAS*. 1217252110

Gautvik, K.M., de Lecea, L., Gautvik, V.T. et al (1996) "Overview of the most prevalent hypothalamus-specific mRNAs, as identified by directional tag PCR subtraction." *PNAS*, vol. 93 no. 16, 8733–8738.

Gellatly, Angus & Zarate, O. (2005) *Introducing Mind and Brain*. Royston: Totem Books.

Gerber, J and Thomas Hummel (2013) "Maternal status regulates cortical responses to the body odor of newborns" *Frontiers of. Psychol.*, 05 September 2013 | doi: 10.3389/fpsyg.2013.00597

Gopnik, Alison.(2011) http://www.ted.com/talks/alison_gopnik_what_do_babies_think/transcript?language=en

Hatfield E, Cacioppo JT, Rapson RL. (1993) "Emotional contagion." *Curr Dir Psychol Sci* 1993;2:96–99.

Hatfield, E. Bensman, L. Thorton, P. and Rapson, R. (2014) "New Perspectives on Emotional Contagion: A Review of Classic and Recent Research on Facial Mimicry and Contagion." Univ. of Hawaii. *Interpersona*. Vol.8 no.2.

Honore, Carl. (2005) *In Praise of Slowness: Challenging the Cult of Speed*. New York: Harper One

Huckshorn, Kevin Ann. (2004) *Creating violence free and coercian free mental health treatment.* www.oregon.gov/DHHS/addiction/trauma-policy/reducing-use.ppt

Iacoboni, M. et al. (2004) "Watching social interactions produces dorsomedial prefrontal and medial parietal bold fMRI signal increases compared to a resting baseline." *Neuroimage* 21, 1167–1173

Janata, Petr. (2009) "The neural architecture of music-evoked autobiographical memories." *Cerebral Cortex* (epub. Feb 24, 2009) 2009 Nov 19 (11) 2579-94.

Janata, Petr. And Birk,J. et al (2002) "The cortical topography of tonal structures underlying Western music." *Science* 2002 Dec 13:298 (5601): 2167-70. DOI: 10.1126/science.1076262

Jee-Yon Lee and Duk-Chul Lee. (2014) "Cardio and Pulmonary Benefits of Forest Walking versus City Walking in Elderly Women: A Randomized, Controlled, Open-Label Trial," *European Journal of Integrative Medicine* (2014): 5-11.

Kandhalu, Preethi. (2013) "Effects of cortisol on physical and psychological aspects of the body and effective ways by which one can reduce stress." *Berkeley Scientific Journal* • Stress • Fall 2013 Volume 18 • Issue 1, 14-16.

Kawakami K. et . (2004) "Effects of phytoncides on blood pressure under restraint stress in SHRSP." *Clin.Exp.Pharmacol.Physiol*. 2004 Dec.31 Suppl.2:S27-8.

Keeler, Jason R. et al. (2015) "The neurochemisty and social flow of singing: bonding and oxytocin." Http://doi.org/10.3389/fnhum.2015.00518

Kimmerer, Robin Wall. (2013) *Braiding Sweetgrass*. Minneapolis: Milkweed Publishers. pg. 236.

Kolb, Bryan. Mychasiuk, Richelle et al. (2012) Experience and the developing prefrontal cortex. PNAS Oct.16, 2012 vol.109 supl. 2 p.17186 – 17193.

Kosfeld, M., Heinrichs, M.et al. (2005) "Oxytocin increases trust in humans." *Nature* 2005 Jun 2;435(7042):673-6.

Kreutz; and Keeler, Jason R. et al (2015) "The neurochemistry and social flow of singing: bonding and oxytocin" *Front. Hum. Neurosci.,* 23 September 2015
| https://doi.org/10.3389/fnhum.2015.00518

Kuhl, Patricia K. (2010) "Brain Mechanisms in Early Language Acquisition." *Neuron.* 2010 Sep 9; 67(5): 713–727. doi: 10.1016/j.neuron.2010.08.038 PMCID: PMC2947444 NIHMSID: NIHMS2343566

Kuhl PK. (2007) "Is speech learning 'gated' by the social brain? *Dev Sci.* 2007 Jan; 10(1):110-20.

Kuhn, Merrily. (2013) "Understanding the Gut Brain: Stress, Appetite, Digestion & Mood." www.ibpceu.com

Kuhn, Simone et al. (2017) "In search of features that constitute an "enriched environment" in humans: Associations between geographical properties and brain structure" *Scientific Reports* 7, Article number: 11920 (2017)

Kreutz, G. and Bongard S. et al (2004) "Effects of choir singing or listening on secretory immunoglobulin A, cortisol, and emotional state." *J Behav Med.* 2004 Dec;27(6):623-35. https://www.ncbi.nlm.nih.gov/pubmed/15669447

Li Q. (2009) "Effect of forest bathing trips on human immune function." *Environ. Health Prev Med. 2010* www.ncbi.nlm.nih.gov/pubmed/19568839

Li Q et al. (2006) "Phytoncides, the essential oils in the bark and stems, induce human natural killer cell activity." *Immunopharmacol Immunotoxicol* Immunopharmacol Immunotoxicol. 2006;28(2):319-33. Source: Department of Hygiene and Public Health, Nippon Medical School, Tokyo, Japan. qing-li@nms.ac.jp

Limb CJ, Braun AR (2008) "Neural Substrates of Spontaneous Musical Performance: An fMRI Study of Jazz Improvisation." *PLoS ONE* 3(2): e1679. doi:10.1371/journal.pone.0001679

Limb, Charles. (2008): "Figure 3. Three-dimensional surface projection of activations and deactivations associated with improvisation during the Jazz paradigm." *PLoS ONE* doi:10.1371/journal.pone.0001679.g003.

Lipton, Bruce. (2005) *The Biology of Belief: Unleashing the power of consciousness, matter & miracles*. Santa Rosa: Elite Books.

REFERENCES

Lundstrom, Johan, Annegret Mathe[4], Benoist Schaal[4,5], Johannes Frasnelli[6], Katharina Nitzsche[7], (2013) "Maternal status regulates cortical responses to the body odor of newborns" *Front Psychol.* 2013; 4: 597. doi: 10.3389/fpsyg.2013.0059 PMCID: PMC3763193 PMID: 24046759

Lyness, D'Arcy (2012) "Encouraging Your Child's Sense of Humor". Http://kidshealth.org/ (1995-2015) Reviewed by D'Arcy Lyness, PhD

Mate, Gabor. (2010) *In the Realm of Hungry Ghosts.* Berkeley: North Atlantic Books

McGilchrist, Iain (2012-07-15). *The Divided Brain and the Search for Meaning* (Kindle Locations 262-266). Yale University Press. Kindle Edition.

Moss M, Oliver L. (2012) "Plasma 1,8-cineole correlates with cognitive performance following exposure to rosemary essential oil aroma." Ther Adv Psychopharmacol. 2012 Jun;2(3):103-13. Doi: 10.1177/2045125312436573.

Osman, S. E., Tischler, V., & Schneider, J. (2016). " 'Singing for the Brain': A qualitative study exploring the health and well-being benefits of singing for people with dementia and their carers." *Dementia, 15*(6), 1326–1339. https://doi.org/10.1177/1471301214556291

Ozdemur et al. (2006) "Singing uses larger bihemispheric network than does speaking: Shared and distinct neural correlates of singing and speaking." *Neuroimage.* 2006, Nov.1 33(2):628-35. Pub Med.

Panksepp, Jaak. (1998) *Affective Neuroscience: The Foundations of Human and Animal Emotions.* New York: Oxford Press

Panksepp, Jaak. & Biven, Lucy. (2012) *The Archeology of Mind: Neuroevolutionary Origins of Human Emotions.* New York: Norton & Co. p.308.

Park, BJ, Tsunetsugu, Y. et al. (2010) "The physiological effects of Shinrin-yoku (taking in forest atmosphere or forest bathing): evidence from field experiments in 24 forests across Japan." *Environ Health Prev. Med.* 2010 Jan 15(1).18-26.

Perry, B. & Pollard, R. (1998) "Homeostasis, Stress.Trauma, and Adaptation: A neurodevelopmental view of childhood trauma." *Child and Adolescent Clinics of North America 7* (1) (January 1998):33-51. Citing data from R. Shore, (1997) *Rethinking the Brain: New Insights into Early Development.* New York: Families and Work Institute.

Porges, Stephen. (2004) Neuroception: A subconscious system for detecting threats and safety. *Zero to Three.* May, 19-24.

Porges, S. (2007) The polyvagal perspective *Biological psychology* 74 (2), 116-143

Porges, Stephen. (2007) Brainstem nuclei that are oxytocin sensitive. Love or Trauma? DVD

Porges, Stephen. (2008) *Love or Trauma?*

Porges, Stephen. (2011) *The Polyvagal Theory: Neurophysical Foundations of Emotions, Attachment, Communication, Self-Regulation.* New York: Norton.

Porges, Stephen. (2013) *The Polyvagal Theory: New Conceptualizations & Clinical Applications.* Seminar

Porges, Stephen. (2017) *The Pocket Guide to the Polyvagal Theory: The Transformative Power of Feeling.* New York: Norton

Porges, Stephen. & Dana D. (2018) *Clinical Applications of the Polyvagal Theory: The Emergence of Polyvagal-Informed Therapies.* New York: W.W. Norton.

Potkin, Katya Trudeau. and Bunney, William E. (2012) "Sleep Improves Memory: The Effect of Sleep on Long Term Memory". *PLOS.* https://doi.org/10.1371/journal.pone.0042191

Provine, Robert R. (2001) *Laughter: A Scientific Investigation.* New York: Penguin

Russo, MA., Santarelli, DM., O'Rouke, D. (2017) "The physiological effects of slow breathing in the healthy human." *Breathe,* 2017, 13.. 298-309

Schladt, T. Moritz et al. (2017) "Choir versus Solo Singing: Effects on Mood, and Salivary Oxytocin and Cortisol Concentrations" Front Hum Neurosci. 2017; 11: 430. Published online 2017 Sep 14. doi: [10.3389/fnhum.2017.00430]PMCID: PMC5603757PMID: 28959197

Shore, R. (1997) *Rethinking the Brain: New Insights into Early Development.* New York: Families and Work Institute.

Shusterman, D.J., Jansen,K., Weaver E. and Koenig J. (2007) "Documentation of the nasal nitric oxide response to humming: methods evaluation." *European Journal of Clinical Investigation* (2007) 37, 746–752 DOI: 10.1111/j.1365-2362.2007.01845.

Siegel, Daniel J. (2007) *The Mindful Brain.* New York: Norton.

Siegel, Daniel J. (2012) *Pocket Guide to Interpersonal Neurobiology.* New York: Norton.

Siegel, D. J.& Bryson, T.P. (2012) *The Whole-Brain Child.* New York: Bantam.

Shonkoff, J.P. (2017). "Rethinking the Definition of Evidence-Based Interventions to Promote Early Childhood Development." *Pediatrics, 140*(6), e20173136.

Sousa, David A. (2006) *How the Brain Learns.* Thousand Oaks: Corwin Press.

Sumioka, Hidenobu, Nakae, A. et al (2013) "Huggable communication medium decreases cortisol levels" *Scientific Reports* 3, Article number: 3034 doi:10.1038/srep03034 23 October 2013

Tanaka-Arakawa Megumi M., Matsui, Mea. et al. (2015) Developmental changes in the corpus callosum from infancy to Early adulthood: A structural magnetic resonance imaging study. Plos One 2015. 10(3): e0118760.

Taylor, A.F. & Kuo, F. (2008) "Children with attention-deficit concentrate better after walk in the park." *Journal of Attention Disorders.* 20 (10) April p.1-7.

Uddin, Lucina Q., Iacoboni, M., Lange, C. and Julian Paul Keenan. (2007) "The self and social cognition: the role of cortical midline structures and mirror neurons." *TRENDS in Cognitive Science.* Vol.11 No.4

WNYC's Radio Lab with Jad Abumrad and Robert Krulwich delving into the making of their acclaimed program that melds science, philosophy, and psychology this podcast "short" from November, (2007),episode "Sound as Touch" with Dr. Anne Fernald

You can grow your intelligence. (2002-2013) *Mindset Works Inc.* www.mindsetworks.com (Dweck)

Vickhoff, Bjorn. et al. (2013) "Music structure determines heart rate variability of singers." *Frontiers in Psychology.* 09 July 2013. www. Frontiers in Auditory Cognitive Neuroscience.org.

Wohlleben, and Harmuth, Frank, et al. (2003) *Der sachsische Wald im Dienst der Allgemeinheit* ("The woods of Saxony in the service of the general public") State Forestry Commission of Saxony. www.smul.sachsen.de/sbs/download/Der_sachsische_Wald.pdf.

Wohlleben, Peter. (2015) *The Hidden Life of Trees.* Vancouver/Berkeley: Greystone Books

Xie et al. (2013) "Sleep initiated fluid flux drives metabolite clearance from the adult brain." *Science*, October 18, 2013. DOI:: 10.1126/science.1241224

Yuan, Lin et al (2016) "Oxytocin inhibits...inflammation in microglial cells..." *Journal of Neuroinflammation*

https://developingchild.harvard.edu/science/key-concepts/brain-architecture/#neuron-footnote

http://www.pbs.org/thisemotionallife/topic/humor/benefits-humor

INDEX

INDEX